Breaking Boundaries

SUNY series in Environmental Governance:
Local-Regional-Global Interactions

Peter Stoett and Owen Temby, editors

Breaking Boundaries

Innovative Practices in
Environmental Communication and
Public Participation

Edited by

Kathleen P. Hunt, Gregg B. Walker,
and Stephen P. Depoe

SUNY
PRESS

Published by State University of New York Press, Albany

© 2019 State University of New York

All rights reserved

Printed in the United States of America

For information, contact State University of New York Press, Albany, NY
www.sunypress.edu

Library of Congress Cataloging-in-Publication Data

Names: Hunt, Kathleen P., editor | Walker, Gregg B., editor | Depoe, Stephen P.,
    editor
Title: Breaking boundaries: Innovative practices in environmental communication and
    public participation / Kathleen P. Hunt, Gregg B. Walker, and Stephen P. Depoe,
    editors.
Description: Albany : State University of New York Press, [2019] | Series:
    SUNY series in Environmental Governance: Local-Regional-Global Interactions |
    Includes bibliographical references and index.
Identifiers: LCCN 2019952340 | ISBN 9781438477053 (hardcover : alk. paper) |
    ISBN 9781438477077 (ebook)
Further information is available at the Library of Congress.

10 9 8 7 6 5 4 3 2 1

Contents

SECTION II.
EXPANDING PATHWAYS OF COMMUNITY ENGAGEMENT

Section III.
Enacting Horizons of Civic Technology

Illustrations

Figures

Tables

Foreword

PETER STOETT AND OWEN TEMBY

An implication of the ascendant paradigm to environmental governance—that ecosystems are structured as non-linear complex adaptive (social-ecological) systems—is that effective governance starts at the scale at which these systems are structured. The pathologies of one-size-fits-all command-and-control policy are well known to scholars and practitioners alike. Our approaches to governing the environment must complexify to manage social-ecosystem complexity and must embrace reflexivity and adapt in response to new knowledge about the environment and stakeholder goals and interests. Successful stakeholder engagement is central to the process in numerous ways: it is necessary to develop and disseminate better knowledge about human interaction with the environment, generate stakeholder buy-in, mollify resistance to important initiatives and, in many cases, satisfy mandated public consultation requirements.

Experienced environmental managers are well aware of the need for public participation, especially in developing productive collaborative processes with stakeholders, and typically seek to facilitate it. However, public engagement is costly and the barriers are plentiful. Social scientists have generally done an inadequate job of providing environmental practitioners with tools to improve the areas in need, like trust development, knowledge coproduction, and shared problem definitions. Knowledge of the relevant globally linked local contexts, and broader lessons that can be adapted to local social-ecological contexts, is needed to guide decision makers. It will be a great challenge to provide, and it means that social scientists have a lot of work to do.

For this reason, *Breaking Boundaries: Innovative Practices in Environmental Communication and Public Participation*, is a welcome contribution. This collection, edited by Kathleen P. Hunt, Gregg B. Walker, and Stephen P. Depoe, provides a state-of-the-art overview of public participation theory and practice in environmental governance from the leading scholars on the topic. The broad set of cases and contexts give the reader a sense of the many different ways in which the public engages with policy makers and government officials (and, in doing so, "breaking boundaries"), including the more disruptive communication practices being used in our era of social media and climate science denial. The collection's valuable lessons (about the importance of allowing for alternative forms of public engagement, of adapting participation infrastructure to the local context, of accounting for power differentials, and of the potential uses and limitations of digital media, to name a few) should inform the work of environmental social scientists seeking to develop the tools enabling government to communicate productively with the public.

We are pleased to include *Breaking Boundaries* in the SUNY Series in Environmental Governance: Local-Regional-Global Interactions and believe that it will become recognized as a valuable contribution to the emerging literature on environmental communication, public engagement and, more broadly, social-ecological systems governance.

Introduction

From Public Participation to Community Engagement—and Beyond

KATHLEEN P. HUNT, SUSAN SENECAH,
GREGG B. WALKER, AND STEPHEN P. DEPOE

This volume starts with the assumption that effective participation by members of a community in policy decisions that impact their lives and livelihoods should be promoted, both as part of a normative commitment to deliberative democracy and because such participation often results in superior and more politically sustainable solutions to thorny or contentious issues (Dryzek, 2010; Fishin, 2011; Guttman & Thompson, 2004). The role of community input in environmental policy making has been a prominent issue for decades, dating back to Rachel Carson's call for a citizen's right to know about exposure to poisons in *Silent Spring* (1962) and the passage of landmark legislation such as the National Environmental Policy Act (NEPA) of 1969, a law that pioneered the creation of explicit pathways for soliciting public comments on proposed governmental actions. Public interest and involvement in environment quality has accelerated in the United States and elsewhere ever since, spurred by the activism and energy of Earth Day in 1970, opposition to nuclear energy, the crisis of hazardous waste disposal, environmental justice concerns, sustainable food and energy choices, climate change mitigation and adaptation, and many other local and global issues (Gottlieb, 1993).

Environmental problems that prompt the consideration of legal remedies or policy options are often multifaceted and complex, and present challenges for effective community involvement. For instance, in many situations, the "perceptions of problems (e.g. the nature of risk and priorities for collective action)" held by scientific or technical experts "are judged to be more rational that the 'subjective' perceptions . . . of the public" (Fiorino, 1990, p. 229), leading to decisions that are less than optimal for impacted stakeholders. Reconciling the need for expertise in managing and protecting the environment with the ideals of participatory democracy remains contentious but is nevertheless critical to the legitimacy of our political system, particularly with respect to the energy-food–natural resource management–environmental protection sectors.

Scholars from a variety of disciplines have generated a substantial body of theories, models, and concepts that inform the design, implementation, and evaluation of participatory processes attached to environmental and related policy arenas (Daniels & Walker, 2001; Nabatachi & Leighninger, 2015). However, effective public engagement remains difficult to do well, illustrating for us the vital need for more meaningful interactions among community members, civic leaders, and other interested parties. With a focus on the relationship between structure and enactment via discourse, the field of environmental communication offers theories and concepts that grapple with questions concerning the value and impact of public influence in environmental decision making, the role of power in the development and implementation of environmental politics, and also what equitable governance can look like (Carvalho, Phillips, & Doyle, 2012; Cox & Pezzullo, 2016; Depoe et al., 2004).

For example, Senecah's (2004) Trinity of Voice (TOV) heuristic articulates three critical conditions that must be met to optimize effective public participation: access (opportunity for expression), standing (civic legitimacy), and influence (equal consideration of perspectives) (pp. 23–25). As a "practical theory," TOV continues to be mobilized in contexts ranging from policy and planning (Carvalho, Pinto-Coelho, & Seixas, 2016; Hall, Gilbertz, Anderson, Lucas, & Ward, 2016; Natarajan, 2017), to natural resource management (Chowdhury & Rahman, 2008; Egunyu & Reed, 2015), to human ecology (Walker, Senecah, & Daniels, 2006). Designing and implementing participatory processes in ways that foreground these conditions is central to building trust between community members and their civic leaders, [building] capacity, and creating better environmental decisions (Senecah, 2004, p. 23).

As some observers have noted, in recent years "the participatory agenda has started to lose its momentum and justification" (Wesselink et al., 2011, p. 2688), and the environmental public sphere seems increasingly under attack. For example, 2016 was an especially turbulent year that featured, among other things, a surprising Brexit electoral result in the United Kingdom, reactionary rumblings across Europe and the Global South, a protracted and at times violent protest by indigenous water protectors opposing the construction of the Dakota Access Pipeline through native lands in North Dakota, and the disturbing election of Donald Trump as president of the United States. Rollbacks in U.S. environmental regulations and withdrawal from international climate change agreements will certainly damage natural systems, have already garnered harsh criticism from global leaders and environmental activists, and will stifle opportunities for expansive public input.

Yet people *are* engaging decision makers in innovative ways, using tactics such as leadership capacity building, protest, and social media campaigns in efforts to reclaim access, civic standing, and influence. Community engagement can take many forms to empower communities and enact discursive change. With Wesselink and colleagues (2011), we call for a "more reflexive awareness of the different ways in which participation is defined and practiced" (p. 2688). Following this, our volume aims to break boundaries—that is, to widen the scope of concepts, practices, and theoretical insights included under this umbrella term. It may not (always) be the case that public participation is waning, but rather that there are other possibilities for public involvement in particular regulatory outcomes or for building awareness and relationships among stakeholders in long(er) term fights for change, as we illustrate below.

Recent Developments—and Lessons Learned

Two recent developments illustrate the contemporary landscape of environmental civic engagement and deliberative democracy; their contrast reinforces the significance of this edited volume. The first represents limitations of what Nabatachi and Leighninger (2015) refer to as the "public participation infrastructure," a concept that informs the traditional view of public participation in environmental decision making. In late 2016, the U.S. Department of Interior Bureau of Land Management (BLM) released a new policy designed to "democratize" the BLM resource management planning process by "making it more collaborative, inclusive, transparent and reflective of landscape-wide priorities" (Shogren, 2016). Although BLM planning decisions are subject to federal public participation requirements, some stakeholders claim that management plans have been drafted behind closed doors, often with minimal opportunities for public feedback offered too late in the process (McElroy, 2014). Titled "Planning 2.0," the new rule called for elevated public participation, providing greater agency-public interaction through shared documents and data, written rationales for resource management plans, as well as public comment periods (Shogren, 2016). Applying Senecah's (2004) TOV framework, Planning 2.0 showed great potential for increasing citizen engagement by supporting the access, standing, and influence of those involved in BLM planning decisions.

In spite of the BLM's intentions, President Obama's original executive order was derided as government overreach (Anderson, 2017). Through strategic deployment of obscure legislative procedures and hidden influence of industry lobby groups (Shogren, 2016), Congress voted to "strip the Bureau of Land Management's (BLM) 'Planning 2.0' rule from the books" (Henry, 2017). The bill was subsequently signed by President Trump on March 27, 2017. In this process, our elected officials communicated their skepticism about (or perhaps disdain for) meaningful and frequent public participation in environmental management and planning, significantly diminishing TOV.

The second development was more widely publicized and, unlike Planning 2.0, occurred *outside* conventional spaces of public policy deliberation. Throughout the spring of 2017, several large-scale public protests were held in Washington, DC, and internationally, mobilizing millions in response to policy retrenchments instigated by the Trump administration. On April 29, 2017, the 100th day of Trump's presidency, several hundred thousand participants marched to the White House under the slogan "There is no Planet B" (People's Climate, 2017). Just one week prior, on Earth Day, the March for Science also brought more than one million people to the National Mall, marching "to defend the role of science in policy and society" (March for Science, 2017). Indeed, the Women's March, convened in January, "was the largest coordinated protest in U.S. history and one of the largest in world history" (Women's March, 2017); participation in this event was nearly three times higher than attendance at the Presidential Inauguration the day before (Wallace & Parlapiano, 2017).

Largely coordinated through social media outlets including Facebook, Twitter, and Reddit, these day-long events not only invited participants to march on Washington, but also sparked satellite protests in hundreds of cities across the United States and around the world (Levinson, 2017; March for Science, 2017; Tamkin & Gramer, 2017). Through activities like crafting hats and making signs, giving speeches and playing music, delivering performances and parodies, posting pictures, and sharing videos from the events, protesters engaged in various creative and innovative tactics to voice their concerns and advocate for social change (Borovic, 2017; Levinson, 2017; People's Climate, 2017). The legacy of these marches endures as their Facebook pages and hashtags remain in circulation, and each of the charter organizations now outline explicit agendas for growing its respective movement. In addition to signaling the emergence of renewed political mobilization, these marches represent "new forms of participation" that become necessary when people "lack control over social decisions that affect them" (Fiorino, 1990, p. 228). Indeed, protest can be an important means by which "stakeholders who are denied [Trinity of Voice] . . . find a means by which to claim them" (Senecah, 2004, p. 23).

We draw two lessons from these developments. First, as the demise of Planning 2.0 illustrates, traditional public participation structures and pathways too often remain insufficiently open, transparent, or fair (Depoe and Delicath, 2004). Indeed, Nabatchi and Leighninger (2015) suggest that "the infrastructure of the public square"—what they conceive as the antiquated laws, processes, and institutions that no longer adequately support robust public engagement—is failing. The prescribed rules and procedures for public hearings, for example, can be difficult for ordinary citizens who may be unaccustomed to making public comments, thereby limiting engagement between decision makers and the public (Cox & Pezzullo, 2016, p. 303). Although policy frameworks such as NEPA often require public comment or other feedback, this can be perceived as perfunctory when regulatory authorities (ab)use public participation processes to "decide, announce, and defend" decisions made a priori rather than collect and incorporate input from affected publics (Hendry, 2004). Amid polarizing environmental policy debates, even the perception that "systems [are] designed to protect [officials'] expertise from citizen interference" (Nabatachi & Leighninger, 2015, p. 3) can sow the seeds of distrust between leaders and community members.

Second, as illustrated in the huge protest marches that unfolded in early 2017, public involvement in environmental decision making is increasingly going beyond traditional channels and mechanisms of participation. To be sure, comment periods, advisory boards, and hearings are still important means by which citizens communicate with decision makers, provide feedback on policies and practices, and voice their concerns. However, protests like the March for Science and People's Climate March demonstrate how publics also use unofficial, local, and disruptive means of public participation.

Mobilized marches and other activities analyzed in this volume support recent findings in the environmental policy literature, where scholars have noted shifts in public participation frameworks as traditional approaches are challenged and innovative alternatives are pursued. Charting this discursive turn over the past decade, Ross and colleagues (2016) find that "worldwide, the term 'public participation' is in decline, while 'community engagement' is rising" (p. 123). While traditional participation requirements can be met with passive methods, such as "command and control" public meetings or comment periods, engagement efforts emphasize active approaches that can foster collaboration and reconfigure the pathways by which community members become involved in environmental governance, such as interactive workshops, visualization exercises, community cafes, and participatory strategic planning. Individuals and grassroots groups also actively engage in deliberative processes through online tools, including using social media and apps, to raise awareness, share resources, recruit participants, and even coordinate affiliated events. Indeed, engaged citizens and communities are increasingly becoming "active agents, often taking their own initiatives to protect environments, whether through advocacy or practical action" (Ross et al., 2016, p. 125).

We are chastened as well as inspired by Planning 2.0 and the 2017 marches because they illustrate how engaged publics are using a widening array of creative, impactful, and sometimes multilayered tactics to advocate for environmental change and participate in decision-making processes. The March for Science and the People's Climate March effectively mobilized people in a way not evident since the Civil Rights and anti–Vietnam War movements of decades past. Participants not only flew, bused, or carpooled to Washington, DC, but also hosted knitting and sign-making parties, sang together, and shared photos and videos online. Thus, as "active agents" (Ross et al., 2016), stakeholders may employ methods and tactics that embrace empowerment—participation that influences people and outcomes—involvement that makes a difference.

"Breaking Boundaries" as our Central Motif

By attending to the range of *and* interactions among "activities by which publics' concerns, needs, interests, and values are incorporated into decisions and actions on public matters and issues" (Nabatchi & Leighninger, 2015, p. 13), scholars and practitioners may generate both theoretical insights and structural reforms. With this in mind, we invoke "breaking boundaries" as an organizing thematic for the critique, celebration, and expansion of what normatively constitutes public participation, while highlighting efforts to persuade, enact, and resist agendas; (re)define environmental problems; and advocate for solutions.

We collectively explore *how stakeholder voices can be reclaimed in regulatory decision making, what is afforded (and foreclosed) by various practices and tactics used to design and navigate public participation processes, and how communities around the world are innovating new technologies to engage environmental decision makers.* This volume is offered in the hope that scholars and practitioners can break the boundaries that demarcate formal public involvement from other civic activities and work across concepts and categories, to not only expose their limits but also productively nuance the interrelationships among them.

Chapter Organization and Cross-Cutting Themes

Following an opening chapter that provides an overview of the public participation innovation landscape, this volume is organized into three sections. Section I, "Exploring Dimensions of Participation Within Policy Frameworks" (chapters 2–6), presents a variety of case studies drawn from traditional public participation infrastructures. Section II, "Expanding Pathways of Community Engagement" (chapters 7–10), explores ways to move engaged publics beyond the limitations of traditional participation toward broader, more inclusive engagement. Section III, "Enacting Horizons of Civic Technology" (chapters 11 and 12), explores the emerging use of information and communication technologies (ICTs) in environmental decision making. As you read this volume, we invite you to extract both theoretical and critical insights and more specific, practical guidance that can improve environmental public participation at all levels. In particular, we want to preview four cross-cutting themes that emerge in the chapters.

1. Public participation infrastructures should adapt to local conditions and stakeholder concerns whenever possible to yield optimal results. Local context should be strongly considered when designing and implementing participation mechanisms, whether they are mandated by regulatory

frameworks or are part of the ongoing fabric of the community. Based
on a variety of local experiences and interventions, Silka et al.
(chapter 1) warn against a "one size fits all" approach to public participation, and
indicate that design choices should be sensitive to issues of scale, range of
stakeholders impacted, and decision complexity. Walker et al. (chapter 2)
provide a number of concrete suggestions for public participation practice
derived from listening sessions with local stakeholders within the U.S. forest
system. Lind (chapter 4) suggests that environmental policy makers should
take into account variations in local geography and politics when designing programs to solicit public input on complex issues such as water use
and quality. Reinig and Sprain (chapter 5) examine cultural discourses in
public participation around energy-related issues in Boulder, Colorado, and
suggest that conventional tools like public hearings, email communications,
and referendums can have different meanings and effectiveness depending
on local norms and patterns of interaction. Dodson and Paliser (chapter
7) compare and contrast two public participation frameworks in New
Zealand and present ideas for improving multiparty dialogue that involve
indigenous voices and expertise in ways that can yield superior and more
supportable decisions based on community access, standing, and influence
as outlined in Senecah's (2004) TOV. Based on her lived experiences and
interviews with community members living in Indonesia, Tam (chapter
8) recommends that public participation planners account for the spatial
particularity of participation practice, meaning that stakeholder input is
sometimes best obtained in byways and neighborhood spaces outside the
government hearing room or town hall.

**2. Innovative public participation practices should combine elements
of traditional infrastructure with alternative formats and approaches to
community engagement.** As indicated earlier, breaking boundaries does not
necessarily mean discarding long-standing mechanisms for public participation such as hearings, comment periods, and advisory boards. Rather, we
see innovations occurring in many instances that involve *combining* old and
new ways of generating community interest and involvement. McKinney
(chapter 3) outlines a regional approach to natural resource management in
the Crown of the Continent region of the northwestern United States and
southwestern Canada. What McKinney terms the "ecology of governance"
in the region involves a dense network of relationships and activities among
a host of governmental and quasigovernmental bodies that encourages a

significant amount of local stakeholder engagement. Upton et al. (chapter 9) outline a program of indigenous leadership development offered in a number of countries by the nonprofit organization Rare, and conclude that the program has produced a variety of benefits including an increase in the amount of local community engagement in more conventional structures of environmental problem solving and governance.

3. While digital media and other information and communication technologies (ICTs) can enhance the quantity and impact of community engagement in environmental decision making, their affordances and limitations should be also be considered. A number of chapters examine the use by community members of so-called "civic technologies" as a way of both gaining information about environmental issues and influencing future outcomes in environmental decision making. Quiring (chapter 11) presents "trans-media storymaking" as an innovative mode of interactive social media engagement that can provide users with information provided by government officials and others as well as ways of inputting ideas and comments into an ongoing discernment process. Typhina (chapter 12) presents "eco-apps" as a technology that can afford users with opportunities for tagging, posting and commenting, and visualization related to local geographic contexts and concerns. Both chapters present preliminary user experience results, with Quiring focusing on counting views of the transmedia website and Typhina focusing on the limitations of various design elements of the eco-apps. While these case studies highlight the civic potential of digital modes of community engagement, they also illustrate that digital technologies may stimulate a variety of forms of engagement, but should not be taken as substitutes for other participation modes that provide access, standing, and influence for stakeholders. As noted by Dubow (2017), "it remains to be seen whether online activism actually translates into positive change in the offline world, and whether the increasing use of digital technologies actually facilitates or hinders greater social inclusion" (p. 2). Thus, governmental entities and others that employ these technologies should account for their limitations as well as their affordances.

4. Public participation formats and technologies should account for power dynamics present in local decision-making contexts and embedded in the formats themselves. The relationship between public participation infrastructures and citizen power is a central element of Trinity of Voice.

For Senecah (2004), voice is a normative concept that is assessed based on the degree to which participation processes provide opportunities for community members and other stakeholders to have a demonstrative impact on the outcome of an environmental policy decision that impacts their lives and livelihoods. A number of the chapters in this volume examine ways in which community members can reclaim political influence and power to shape issues and decisions, either within or outside official participation channels. Innovative or creative tactics may either disrupt or transform conventional public participation strategies. Hunt et al. (chapter 6) examine ways in which participants in public hearings may subvert institutional efforts to label activist discourse as "indecorous voice" in order to interrupt routinized processes in ways that promote maximum visibility for otherwise forgotten issues or communities. Simis-Wilkinson and Hopke (chapter 9) coin the term "disruptive public participation" to describe how activists can employ various media platforms (like Twitter) and forms (words and images) to create and circulate a counternarrative that critiques institutional rules and use of force in ongoing environmental protests both within and outside the meeting hall. Innovative strategies may either disrupt or transform conventional approaches to public participation.

The landscape of public engagement is changing as the environmental public sphere is under threat; traditional ways of communicating and enacting public participation in environmental decision making must be reclaimed, reinvigorated, and transformed. As revealed in this volume, innovations taking place in local communities can nourish a new generation of deliberative engagement in the environmental arena. We invite readers to reconsider the conventions of public participation in environmental decision making amid the realities of 21st-century governance, and imagine possibilities for new processes and structures for citizen engagement in the United States and around the world.

References

Anderson, C. (2017). Congress should approve BLM's "Planning 2.0" initiative. *The Hill*. Retrieved from http://origin-nyi.thehill.com/blogs/pundits-blog/energy-environment/321874-congress-should-approve-blms-planning-20-initiative

Borovic, K. (2017, January 21). What are the pink hats at the Women's March? This accessory speaks volumes. *Bustle*. Retrieved from https://www.bustle.com/p/what-are-the-pink-hats-at-the-womens-march-this-accessory-speaks-volumes-32048

Bureau of Land Management (BLM). (2017). What we manage. Retrieved from https://www.blm.gov/about/what-we-manage/national

Carvalho, A., Pinto-Coelho, Z., & Seixas, E. (2016). Listening to the public—enacting power: Citizen access, standing and influence in public participation discourses. *Journal of Environmental Policy & Planning*. Advance online publication. doi: 10.1080/1523908X.2016.1149772

Carvalho, A., Phillips, L., & Doyle, J. (Eds). (2012). *Citizen voices: Performing public participation in science and environment communication*. Bristol, UK: Intellect.

Chowdhury, R., & Rahman, R. (2008). Multicriteria decision analysis in water resources management: The malnichara channel improvement. *International Journal of Environment Science and Technology, 5*(2), 195–204. doi:10.1007/BF03326013.

Cox, J. R., & Pezzullo, P. C. (2016). *Environmental communication in the public sphere* (4th ed.). Los Angeles, CA: Sage Publications.

Creighton, J. L. (2005). *The public participation handbook: Making better decisions through citizen involvement*. San Francisco, CA: Jossey-Bass.

Daniels, S. E., & Walker, G. B. (2001). *Working through environmental conflict: The collaborative learning approach*. Westport, CT: Praeger.

Depoe, S., & Delicath, J. W. (2004). Introduction. In S. P. Depoe, J. W. Delicath, & M. A. Elsenbeer (Eds.), *Communication and public participation in environmental decision making* (pp. 1–10). Albany, NY: State University of New York Press.

Dryzek, J. S. (Ed.). (2010). *Foundations and frontiers of deliberative democracy*. Oxford, UK: Oxford University Press.

Dubow, T. (2017). *Civic engagement: How can digital technologies underpin citizen-powered democracy?* Santa Monica, CA: Rand Corporation.

Egunyu, F., & Reed, M. (2015). Social learning by whom? Assessing gendered opportunities for participation and social learning in collaborative forest governance. *Ecology and Society, 20*. doi:10.5751/ES-08126-200444.

Fishin, J. S. (2011). *When the people are talking: Deliberative democracy and public consultation*. Oxford, UK: Oxford University Press.

Fiorino, D. J. (1990). Citizen participation and environmental risk: A survey of institutional mechanisms. *Technology and Human Values, 15*(2), 226–243.

Gottlieb, R. (1993). *Forcing the spring: The transformation of the American environmental movement*. Washington, DC: Island Press.

Guttman, A., & Thompson, D. (2004). *Why deliberative democracy?* Princeton, NJ: Princeton University Press.

Hall, D., Gilbertz, S., Anderson, M. B., Lucas, C., & Ward, L. (2016). Beyond "buy-in": Designing citizen participation in water planning as research. *Journal of Cleaner Production, 133*, 725–734.

Hendry, J. (2004). Decide, announce, defend: Turning the NEPA process into an advocacy tool rather than a decision-making tool. In S. P. Depoe, J. W. Delicath, & M.-F. A. Elsenbeer (Eds.), *Communication and public participation in environmental decision making* (pp. 99–112). Albany, NY: State University of New York Press.

Henry, D. (2017, 7 February). House passes bill to block Obama land planning rule. *The Hill.* Retrieved from http://thehill.com/policy/energy-environment/318372-house-passes-bill-to-block-obama-land-planning-rule

International Association of Public Participation (IAP2). (2014). IAP2's Spectrum of Public Participation. Retrieved from https://c.ymcdn.com/sites/www.iap2.org/resource/resmgr/foundations_course/IAP2_P2_Spectrum_FINAL.pdf

Levinson, E. (2017, April 29). Climate protest takes on Trump's policies—and the heat—in DC march. Retrieved from http://www.cnn.com/2017/04/29/us/climate-change-march/index.html

March for Science (2017). Why we march. Retrieved from https://www.marchfor-science.com/why-we-march/

McElroy, W. (2014, March 27). A vast land grab to grouse about. *The Daily Bell.* Retrieved from http://www.thedailybell.com/editorials/wendy-mcelroy-a-vast-land-grab-to-grouse-about/

Nabatchi, T., & Leighninger, M. (2015). *Public participation for 21st century democracy.* Hoboken, NJ: Wiley.

Natarajan, L. (2017). Socio-spatial learning: A case study of community knowledge in participatory spatial planning. *Progress in Planning, 111,* 1–23.

People's Climate March (2017). Homepage. Retrieved from https://peoplesclimate.org/

Ross, H., Baldwin, C., & Carter, R. W. (2016). Subtle implications: public participation versus community engagement in environmental decision-making. *Australasian Journal of Environmental Management, 23*(2), 123–129.

Senecah, S. L. (2004). The trinity of voice: The Role of practical theory in planning and evaluating the effectiveness of environmental participatory processes. In S. P. Depoe, J. W. Delicath, & M.-F. A. Elsenbeer (Eds.), *Communication and public participation in environmental decision making* (pp. 13–33). Albany, NY: State University of New York Press.

Shogren, E. (2016). How the BLM is overhauling land use planning. *High Country News.* Retrieved from http://www.hcn.org/issues/48.9/how-the-blm-is-overhauling-land-use-planning

Tamkin, E., & Gramer, R. (2017). The Women's March heard round the world. *Foreign Policy.* Retrieved from http://foreignpolicy.com/2017/01/21/the-womens-march-heard-round-the-world/

Walker, G., Senecah, S., & Daniels, S. (2006). From the forest to the river: Citizens' views of stakeholder engagement. *Human Ecology Review, 13*(2), 193–202.

Wallace, T. & Parlapiano, A. (2017). Crowd scientists say women's march in Washington had 3 times as many people as Trump's inauguration. *The New York Times.* Retrieved from https://www.nytimes.com/interactive/2017/01/22/us/politics/womens-march-trump-crowd-estimates.html

Wesselink, A., Paavola, J., Fritsch, O., & Renn, O. (2011). Rationales for public participation in environmental policy and governance: Practitioners' perspectives. *Environment and Planning, 43*(11), 2688–2704.

Women's March. (2017). The march. Retrieved from https://www.womensmarch.com/march

Opening Reflections

Chapter 1

Health, the Environment, and Sustainability

Emergent Communication Lessons
Across Highly Diverse Public Participation Activities

Linda Silka, Bridie McGreavy, and David D. Hart

Focusing on communication and the roles community members should play in decision making is essential to addressing complex sustainability problems related to environmental quality and human well-being, a need increasingly articulated in diverse sites and ways (cf. Cox & Pezzullo, 2016; Moser & Dilling, 2007). For example, the U.S. Environmental Protection Agency (EPA) recently requested proposals that integrate research on ecosystems services and human well-being. This RFP, unlike others, asked for a detailed plan about how community members and project stakeholders can participate in environmental decision making. We see this again in news coverage about the 2015 Paris climate talks, which describe how the development of a transparent and credible decision-making process for carbon emissions reduction was fundamental to the purported success (Davenport, Gillis, Chan, & Eddy, 2015). And, as a final example, the Environmental Communication (EC) List curated by the International Environmental Communication Association (IECA), dated May 13, 2016, includes a series of posts describing the need to bring communication to bear on complex issues including climate change, nuclear power development, ocean acidification, tribal justice, and dryland management.

Yet one simple juxtaposition in the above series of posts from 2016 illustrates a core challenge in efforts to generalize from diverse cases about effective communication strategies to support public participation (Wolf & Moser, 2011). In the first post, the author describes how we must find common ground with those who may not share our views about climate change. Immediately following in the next post, the authors advocate that we must find ways to change minds and behavior related to climate change. This single juxtaposition of best practices related to climate change communication illustrates a basic tension that we engage: 1) how do we *generalize* about communication strategies when the topics that we study, our approaches to communication, and the context-dependent strategies for effective communication are so diverse and sometimes contradictory; and in our efforts to generalize, 2) how do we *avoid getting stuck* and remain ready to incorporate new ideas as we encounter them? Both of these challenges—with generalizability and with stuckness (what psychologists refer to as functional fixedness [Duncker, 1945; McCaffrey, 2012])—are partly the outgrowth of underlying academic norms related to knowledge production. These challenges are an inescapable part of life within complex adaptive systems that are characterized by sensitivity to types of difference (Lansing, 2003; Norberg & Cumming, 2008) and where patterns of difference are fundamentally shaped by language (Burke, 1969). However, we have come to recognize that these challenges can be productively engaged using an array of innovation strategies from other fields (Brown, Harris, & Russell, 2010).

The innovation literature suggests ways to engage this tension, as scholars argue that innovations are likely to occur when experiences across different domains are intermingled (Easley & Kleinberg, 2010; Fox & Cooper, 2013; Marrone, 2010). Approaching problems in fresh ways requires that we devote efforts to strengthening the practice of learning across differences (Brzustowski, 2012; Silka, Kelly, & Ward, 2014; Van de Ven, Polley, Garud, & Venkataraman, 1999). Elinor Ostrom (2007), a Nobel Prize–winning sustainability scientist, has also shown how complexity and context dependency require sensitivity to the details of a socioenvironmental system and adaptability over time. One of the strategies she used to grapple with context dependency and generalizability has been the development of frameworks to guide choices about design, analysis, and adaptation. Following Ostrom, we seek to create adaptable public engagement processes. Ostrom and others have also demonstrated the value of engagement processes that promote learning, are inclusive and transparent, and that allow collaborators to reach their goals and seek to build from identified practices (Clark et al. 2011; Daniels & Walker, 2001; Dietz & Stern, 2008).

In this chapter, we illustrate what can emerge from comparing and contrasting across five elements of public participation in environmental decision making: disciplines, topics, decisions, scale, and partners. These elements can serve as an interpretive framework for seeking new insights across differences in projects and contexts. We have used these elements to help identify factors that influence our myriad collaborations. Our public engagement efforts span issues with a combined focus on the environment, human health and well-being, and sustainability as the fundamental integration of environment and human well-being. We intentionally draw from theory and practice typically seen as outside the discipline of communication to spur new learning and interdisciplinary insights.

With funding from the National Science Foundation (NSF), the National Institutes of Health (NIH), the EPA, and many foundations, we have developed stakeholder-researcher partnerships to increase collaborative research, solve real-world problems, and strengthen decision making for environmental policy at local, state, and federal scales. Much of our work has occurred in sustainability science (Kates et al., 2001; Clark, van Kerkhoff, Lebel, & Gallopin, 2016), and below we highlight exemplary cases drawn from the dozens of projects in which we have been involved in Maine, Massachusetts, New Hampshire, and Rhode Island. We compare and contrast projects aimed at creating a landscape-scale policy for wetland regulation; developing effective risk communication approaches for immigrant communities relying on a contaminated river for sustenance fishing; taking a systems approach to decision making about dams that balances such goals as fish restoration, energy production, and justice and cultural renewal; linking science and other forms of knowledge with decision making for improved water quality, public health, and shellfishing livelihoods; promoting awareness and action related to arsenic in well water; and creating a statewide solid waste management plan.

The diversity in this partial list underscores the importance of our overarching focus on generalizability and innovation. Although typologies for public participation in addressing sustainability challenges have been developed (e.g., McNie, Parris, & Sarewitz, 2016; Miller, 2015), any sustainability science effort will be characterized by diverse types of involvement, the stage of the project, the types of questions being pursued, funding priorities, and the time frame for research outputs (McGreavy & Hart, 2017). That said, when we refer to public participation in our work, we are generally referring to a set of overlapping public participation contexts, including collaborative networks across multiple academic and nonacademic institutions in which citizens and diverse experts work together; interdisciplinary teams in which faculty and students advance research together, often with some level of participation by decision makers who both shape and use the science; public forums in which science is communicated to inform ongoing decision-making processes; and science policy advisory committees where researchers provide expert guidance on technical or regulatory decisions.

In each context, decision making refers broadly to choices that advance identified policies. These policies are often focused on local, municipal, or state governance contexts, and we also seek to inform organizational and less formal institutional policies to link knowledge with action. For example, the second author (McGreavy) is a member of Maine's Shellfish Advisory Council, where she links communication research with decision making about shellfish management issues, such as how to create communication infrastructure and promote access to information in the yearly Fishermen's Forum. Communication fundamentally shapes efforts to link science and relevant knowledge with informed policy choices. Throughout our work, we have strived to produce generalizations without losing sight of context dependency and the necessary specificity of tailored approaches to public participation in environmental decision making.

In the upcoming sections, we provide examples that illustrate differences across *disciplines, topics, scales, decisions*, and *partners* that represent a recurring challenge for finding generalizable solutions across different situations. Our goal is to make visible the diversity of elements shaping human-environmental situations that require innovative efforts if learning across differences is to occur. We describe examples of tailored communication and collaboration strategies, including structured learning activities, shared writing projects, research-based approaches to designing engagement, decision-support development, and web-based communication tools. We highlight these and other strategies and provide recommendations for adapting these for use in other contexts to help generalize and maintain openness to new insights.

Integrating Disciplinary Differences for Transformative Learning

Innovations often occur not when everyone working on a problem has the same expertise, but when they meld expertise from different disciplines (Van de Ven, Polley, Garud, & Venkataraman, 1999). Indeed, the urgency of environmental problems has led to growing calls to combine disciplinary perspectives (Palmer 2012; Pretty, 2011), and many studies of interdisciplinarity advance the conversation (Graybill et al., 2006; Miller et al., 2008; O'Rourke, Crowley, Eigenbrode, & Wulfhorst, 2013). But interdisciplinary success remains elusive, and efforts to bring disciplines together can even fall flat on the problems to which each discipline has devoted substantial attention (Newell, 1998). Too often interdisciplinary collaborations exact formidable transaction costs with little visible result (Brown, Deletic, & Wang, 2015). So why attempt interdisciplinarity at all? What could make it work? Can it be done differently?

Finding ways to negotiate differences in worldviews and in problem framing has been identified as a key challenge in collaborations (Eigenbrode et al., 2007; Miller et al., 2008). Our approaches to public participation in environmental decision making have taken on new forms as a result of integrating disciplines with an emphasis on incorporating each other's generative concepts, drawing from our expertise in social and community psychology (Silka), environmental communication (McGreavy), ecology (Hart), and sustainability science (All).

As we tackle a concrete problem, we often proceed by drawing generative concepts from each other's disciplines rather than through our respective research approaches. From her discipline, McGreavy brings "boundary spanning" and "use of metaphors." From his discipline, Hart has brought to the conversation "coupled social-ecological systems," "knowledge systems," and "resilience." From her discipline, Silka introduces "community-university research partnerships" and "attitude-behavior links." Our work has been transformed by the form these concepts take when they are brought out of our disciplinary "silos" and into the interdisciplinary arena. As a result, our ideas don't just touch at the corners. They confront and interrogate each other, sparking innovation. For example, the concept of resilience has been important for understanding ecosystems, but has its own body of research in psychology on how individuals become resilient in the face of circumstances that typically create psychological damage (Berkes & Ross, 2013). What does the research indicate about what produces resilience at the individual level? Can this offer insights into what might be important in shaping resilient ecosystems? Can these two different disciplinary formulations help each other? The focus is not on complete interoperability but on generativity. The goal is not to reify a particular discipline's framework but rather to see how putting concepts together can address "wicked problems" (Brown, Harris, & Russell, 2010) in more productive ways.

And the disciplines of communication, ecology, and psychology are by no means the only ones involved in our work: there have been many others (e.g., economics, education, engineering, history, law). Each brings generative concepts that can prompt new ways of looking at familiar problems. Our work has been reshaped by becoming familiar with each other's disciplinary journals (and looking at our own journals through others' disciplinary lenses) and by attending each other's disciplinary gatherings (and looking at our own events differently when our team members from other disciplines attend). The strength of this approach is the way it can reduce the "silos" of our disciplines without devaluing what disciplines bring to the conversation (Silka, 2014). We use this concept-based juxtaposition to organize the remaining sections, intending to spark generalizable insights that maintain openness to creative adaptation in the pursuit of solutions.

Juxtaposing Diverse Topics for New Insights

Environmental topics vary so greatly that it can be hard to envision an approach for learning across variety. A review of studies of public participation in environmental communication brings the diversity of possible topics into relief, as scholars and practitioners focus on how communication structures participation in decision making about climate change (Endres, Sprain, & Peterson, 2008), toxins (Pezzullo, 2007), water pollution (Druschke, 2013), nuclear waste siting (Endres, 2009), social-environmental justice (Sandler & Pezzullo, 2007; Sowards, 2012), and road building and habitat fragmentation (Smith & Norton, 2013). Finding coherence across a diverse set of topics is even more challenging because these problems are "interested" (i.e., they are defined in ways that construct our understanding of an issue and leave out other problem framings; see Cox, 2007). For example, defining water pollution as a problem can be shaped by those who are focused on its causes (e.g., industrial wastewater), whereas others may view the cost to businesses of reducing wastewater discharge as the problem. Further, defining water pollution as an environmental problem is based on a set of assumptions about what is an environmental or social issue (Cox, 2007; Schwarze, 2007), which can reify problematic binaries, preventing us from understanding how water pollution spans social and environmental boundaries.

Each topic is characterized by a unique set of temporal and spatial scales across which the issue occurs and for which decision making could have an impact; the level of complexity and uncertainty in the information required to make informed decisions; and the number, type, and expertise of those who have a stake in the decision. This causes us to ask: How do we identify the relevant information for devising communication and public participation strategies; select the ideas that are usable; and decide which approaches on one topic might work for another? Here we consider two water-related environmental topics: vernal pools and contaminated rivers. Though they both involve water as a topic, juxtaposing them allows us to identify some of the differences we need to account for in seeking generalizable insights. Practitioners often comment on the difficulty of drawing lessons from each other's case studies. If one is trying to design a public participation strategy in the western United States, for example, land ownership patterns (i.e., much of the land is under federal jurisdiction) make it hard to learn from public participation approaches in the eastern United States, where land is often under private ownership. Yet the process of comparing very different case studies can promote innovative problem solving that moves beyond rejecting the differences to systematically analyzing differences and identifying patterns that connect (McGreavy, Druschke, Sprain, Thompson, & Lindenfeld, 2017; Sprain & Timpson, 2012).

Wetland filling has contributed to the loss of approximately 53% of wetland habitat in the United States (Dahl, 1990) and is linked with the precipitous decline in amphibian species globally (Stuart et al., 2004). Efforts to promote economic development can also be significantly constrained by wetlands, so there is a need for approaches that balance environmental, social, and economic goals. Starting in the mid-1990s, an interdisciplinary team led by Aram Calhoun and collaborating researchers, government representatives, nonprofit organizations, business leaders, and citizen scientists have completed extensive research to develop adaptive regulations for vernal pools, a type of wetland that is especially common in Maine (e.g., Calhoun & Reilly, 2008; Calhoun et al., 2014; Jansujwicz, Calhoun, & Lilieholme, 2013; McGreavy, Calhoun, Jansujwicz, & Levesque, 2016). Vernal pools are wetland habitats that remain inundated for a portion of each year. Because of their ephemeral nature, small size, and common occurrence on private lands, their central role in ecological processes is often underestimated. They turn out, however, to be essential habitat for many amphibians. Land use changes such as urbanization can disrupt dispersal routes from the pools to these terrestrial habitats.

In contrast to vernal pools, rivers often play a central role in human affairs, and interactions between rivers and society are manifested via a diverse set of biophysical, socioeconomic, and cultural processes. These interactions can also contribute to environmental problems, including water pollution, fishery depletion, invasive species, and water scarcity (Allan & Castillo, 2007). The Merrimack River, which flows through New Hampshire and Massachusetts on its way to the Atlantic Ocean, has figured in much of the history of the region and has suffered for several centuries from heavy pollution from unregulated industry that was the engine of growth for the regional economy (Malone, 2009; Marion, 2014; Steinberg, 1991). New populations of refugees with strong traditions of sustenance fishing moved into the area and fish the waters of the Merrimack (Pho, Gerson, & Cowan, 2008). Accordingly, the health risks of fish consumption and the importance of avoiding fishing have been a focus of environmental agencies.

Both of these topics focus on water resources, yet they exhibit differences in spatial extent, availability and uncertainty in the information required to make decisions, and the requirements for effective public participation. Attending to these differences and tailoring the public participation so it fits these characteristics was therefore essential. For example, Calhoun and her team began by examining how decisions were made and how various science "gaps" constrained the development of decisions that balanced environmental protection with economic development (Levesque, Calhoun, Bell, & Johnson, 2016). This process revealed that science per se is rarely sufficient to ensure the effective management of vernal pools. In particular, they learned that their work was more likely to be used in decision-making processes when the research: 1) addresses the concerns of stakeholders (Calhoun et al., 2014); 2) has undergone peer review to ensure its technical validity (McGreavy et al., 2016); and 3) is part of a stakeholder participation process that is perceived as inclusive, transparent, and fair (Cash et al., 2003; Hart & Calhoun, 2010). Communication was central to ensuring that stakeholders had access to the fora, sufficient information about potential ecological and economic impacts, and the ability to influence choices in the development of the research and policy (Senecah, 2004; Walker, Senecah, & Daniels, 2006). The consistent attention to communication and effective participation of stakeholders promoted the team's ability to conduct research that resulted in "usable knowledge" (Clark et al., 2016).

In the Merrimack River, the environmental agencies wanted to know how to engage with immigrants, how to assist them in understanding the contamination, and how to reduce fish consumption. Reaching individual families who were fishing as a way to reestablish family traditions from their home countries was a key challenge (Silka, 2002a; Silka, 2002b). Tailored participation strategies were undertaken between community leaders and families, where the community leaders spent time with families learning about fishing traditions and talking about fish consumption strategies they might use that do not threaten environmental health (i.e., avoidance of the fatty parts of fish). Public participation was designed based on this shared learning.

These are but two illustrations of the myriad topics encountered in efforts aimed at public participation and environmental communication (Cox & Pezzullo, 2016; Depoe, Delicath & Elsenbeer, 2004). Separately, each suggests particular ways to go about public participation in environmental decision making. By juxtaposing the topics, one becomes cognizant of the characteristics that are not salient when the focus is on one topic: variations in the nature of the problem that might suggest new possibilities for framing; the spatial and temporal scales of the problem's causes and consequences; the target for change (e.g., formal policy or individual behavior); and the degree to which the focus is interlinked with other issues such as public health or economic development. Reflecting on one in comparison with the other can make new possibilities come to light (Hofstadter & Sander, 2013). It can help us see things that might transcend topics. The same actions cannot be taken in every case, but the topic can be viewed in enlarged ways.

Considering How Scale Dependence Shapes Solutions-Focused Strategies

Environmental problems, and the approaches needed to solve them, often require a consideration of multiple spatial and temporal scales. Entire river basins have sometimes been the focus of attention; at other times, the efforts have been directed at a small river segment. Moreover, some problems, such as heavy metal contamination stemming from industrial activities, have a long history where other problems (e.g., potential spread of an invasive species) have not yet occurred. In a new project focused on the future of dams, we are concerned with how decisions (e.g., to remove or repair an aging mill dam or relicense a hydropower dam) are made at the scale of individual dams as well as across thousands of dams in New England. For example, the removal of a single dam may yield no benefits for the restoration of sea-run fisheries such as salmon if downstream dams remain. Thus, decisions about individual dams can often benefit from multiscale analyses that consider the tradeoffs of different decisions across a network of dams (Owen & Apse, 2014).

The issue of how efforts in small areas can be scaled up to larger areas, and vice versa, has been a focus of sustainability deliberations (Cash et al.,

2006; Gismondi, Connelly, Beckie, Markey, & Roseland, 2016; Kates et al., 2001; Norberg & Cumming, 2008). For example, in the case of arsenic in well water, having access to information about arsenic concentrations in private water sources can help individuals decide about filtration and well replacement. At a community scale, however, publicly posting information about the known sources of contamination for individual households can negatively impact private property values (Boyle et al., 2010). This detail influenced how we collected and shared information in a public participation effort organized around community-based monitoring of well water.

Scale issues thus interact significantly with public participation in environmental decision making. Just as ecological processes vary with scale (Levin, 1992), communication processes also differ. The factors that make for more impactful communication (e.g., shared identities, higher trust, opportunities to personalize message) are in greater abundance at the local level. At the community level, issues can be discussed face-to-face. Yet the levers for change may be outside the community where such communication occurs. Broadening the scale may reveal where there are overlapping jurisdictions (e.g., local, county, state), but these jurisdictions may have contradictory rules. Our Safe Beaches and Shellfish Project, which seeks to inform decision making regarding the causes and consequences of coastal water quality impairment, faced major challenges in bringing together the multiple governance scales that affect decisions about beaches and shellfish beds. Water quality at Maine beaches is voluntarily monitored by a nonprofit organization that receives funding from the EPA's Beaches Act. Water quality in shellfish growing areas is federally regulated by the US Food and Drug Administration (FDA) in partnership with the Interstate Shellfish Sanitation Council, which determines how the Maine Department of Marine Resources tests the water and makes decision to protect public health. Despite the seeming topical and spatial similarities, the jurisdictional differences lead to contrasting decision-making needs and abilities.

When researchers focus on public participation at only one scale, they may devise strategies that might not hold for other scales. By moving back and forth across scales, it becomes more apparent that one size does not fit all, and thus there is a relative need for scale-dependent communication strategies (e.g., social media versus face-to-face). The spatial and temporal scale of a problem partially determine the number and type of people who need to be included if a social-environmental problem is going to be meaningfully addressed.

Acknowledging Differences in Decision-Maker Types, Contexts, and Information Complexity

One goal of environmental communication research is to promote learning for informed decision making (Cox & Depoe, 2015; Daniels & Walker, 2001). Ideally, this research can lead to generalizable approaches that can be implemented across contexts. However, variation in decision scenarios, such as who is making the decision, the nature of the topic, and the type and availability of relevant information, can complicate broader application of insights (Beierle & Cayford, 2002). For decisions about arsenic in drinking water—a very significant problem in a rural state like Maine, where a large percentage of homeowners obtain their water from private wells—it is individual landowners who must ultimately assess their arsenic exposure risk and decide among potential solutions. In contrast, many land use decisions in New England—including those involving vernal pools—are mediated by communication processes between private landowners and municipal governments, with considerable input from state and federal agencies. A region may be trying to develop a framework for development so as to anticipate future environmental issues. State policy makers may be attempting to decide whether to restrict shellfishing areas because of contamination. Decisions about dam removal often must consider issues about fisheries, power production, water supply, and recreational uses, which usually involve many different stakeholders and multiple levels of governance.

In short, the decision contexts differ greatly. It could be individuals versus groups who are faced with making the decisions. The decisions could be about short-term issues or issues that are much longer in their emergence or impact. The decisions could be for a small area or a much larger one. Given all of these differences, how do we arrive at a robust, transferable, generalizable communication approach? Is there anything that can transcend these variations in context?

When different contexts are compared and contrasted, one element that nearly all share is the overwhelming amounts of information that must be brought together in deciding what to do. As Gismondi et al. (2016) note: "How can we respond quickly and effectively to this sustainability change? Information is not enough, that much we know. We cannot just put information in front of decision-makers and wait for them to make the right decisions" (p. 8). Something needs to happen if the communication of large amounts of complex, potentially disparate, and sometimes conflicting information is to be transformed into an effective process.

One promising strategy is the development of decision support systems, which include a wide range of analytical approaches for examining the tradeoffs and uncertainties of alternative decision options (Ahmadi et al., 2015). In our experience, such tools work best when they are designed to reflect the multifaceted nature of real-world problems and decision-making contexts, include diverse stakeholders in the process of development and use, and are based on robust methods to ensure their reliability (Waring, 2012).

Together with our partners, we have developed decision support tools for land use planning, the management of coastal water quality, and renewable energy development. For example, Spencer Meyer and his colleagues have worked with communities, businesses, and citizens to create a decision-support tool called the Maine Futures Community Mapper (http://www.mainelandusefutures.org/), which facilitates the development of strategies to balance the conflicting goals of real estate development, forestry, agriculture, and conservation. (Meyer et al., 2014). In contrast to the tendency for many land use decisions to be made in an uncoordinated manner, this five-year project brought together researchers and more than 75 stakeholders to identify the drivers of land use change and to develop models to predict the suitability of different parcels for these alternative land uses. These models were then incorporated into the Maine Futures Community Mapper, an open-access tool that allows diverse users to identify potential conflicts and compatibilities among different land uses. This tool also helps examine the potential influence of policies that promote environmental protection and/or economic development on future land use patterns.

Although decision-support tools can be helpful, they should not be reified or viewed as a substitute for active engagement with stakeholders (Clark et al., 2011). Indeed, research has demonstrated that such tools can be serve as "boundary objects" that facilitate productive interactions in diverse groups (Guston, 2001; McGreavy, Hutchins, Smith, Lindenfeld, & Silka, 2013). We view decision-support tools as a focal point for innovation, and believe they can help untangle some of the "wickedness" of wicked problems.

Adapting to the Engagement Preferences and Needs of Diverse Partners

Even as we strive to develop overarching rules about effective structure for public participation, we also recognize that communication strategies need to be tailored to those we want to reach and to the particular decisions they will confront. These partners can be individual decision makers, small groups, or an entire community. The partners could be policy makers, business leaders, or elected officials at the local, state, or national level. They could be people who are in agreement with each other or a community in which individuals hold vastly different views. They could be people well versed in the environmental issue in question or those who have not yet learned about the issue. They could be people who have grown up in the area where the environmental problem is making itself felt, or they could be individuals who are new. They could be people who live near each other in urban areas or people who are highly dispersed and unable to talk together face-to-face. Too often what we have learned in working with one type of partner won't work for another.

There is a fundamental conundrum here: *what* can we take from efforts with one kind of audience, and *how* can we apply it to new ones? In some instances, the differences may be so stark as to conclude that nothing learned with one group can be generalized to others. Thus, we resort to starting from scratch. Alternatively, people may simply apply these rules as though they are universally applicable. In the former case, we risk ignoring what has come before and generating new rules for every situation. This strategy is also problematic because of the time required to develop public participation processes. In the latter case, we are adopting guidance that we assume is a good fit but may not be.

Part of the challenge is that often the advice is too general, taking the form of truisms: "Be sure to keep your audience's views in mind"; "Understand where people are starting"; "Keep in mind that individuals behave differently than do groups." Best practices and overarching recommendations can be useful starting points, but they can be set at an altitude that gives a broad view but may miss the fine-grained details. For example, Senecah's (2004) articulation of how access, standing, and influence comprise a Trinity of Voice in natural resource policy contexts provides an important heuristic for studying, designing, and improving public participation. However, what access means or how influence is demonstrated can vary, especially in situations with diverse partners. Similarly, Walker et al. (this volume) found that engaging in listening sessions was essential for the diverse stakeholders who were involved in the forest plan revision process. Although this need to engage in active and iterative listening is an important recommendation, what listening means and how it is realized are often context dependent.

We offer two examples to illustrate how the specific practices of information sharing, listening, and communication varied across our work with new immigrants and our efforts with solid waste partners to underscore the need to engage the tension between generalizability and context dependency. Our work with new immigrants and solid waste stakeholders has underscored the need to consider the strengths and limitations of generalization.

Many communication strategies were implemented with new immigrant communities (Silka, 2002c; Silka, 2007). With EPA funding, a television show was created in which Southeast Asian youth interviewed their Buddhist elders and environmental officials (Chao & Long, 2004; Silka, 2002a). This approach was adopted because Southeast Asian cultures discourage youth from being knowledge holders, yet young people were the ones who were learning the information (Silka, 2002a). This approach was also used in building a communication strategy around a Southeast Asian Water Festival that re-created on the Merrimack the traditions on the Mekong but added environmental themes. Lowell's housing stock is old, with many contaminant concerns (Coppens, Silka, Khakeo, & Benfey, 2000). Other communication strategies emphasized the highly respected communicators in the Southeast Asian community (Grigg-Saito, Leong, Och, Silka, & Toof, 2008). Some communicators turned out to be local Southeast Asian women who operated home daycare centers; they taught families with young children how to recognize and address environmental threats in their homes. Similarly, English as a second language (ESL) courses were a community resource. Many newcomers were taking ESL, and instructors needed content to build lessons around home air quality.

Fishing in the contaminated Merrimack River was an urgent problem, requiring that fish advisories be distributed to Southeast Asian families. It was also important to understand the centrality of fishing and the multiple reasons immigrant newcomers continued to fish. Visiting Buddhist temples provided information about cultural practices and dietary traditions. For example, stories were gathered about favorite fish recipes for the many cultures. The stories and recipes were then combined to illustrate Lowell's shared fishing traditions; these stories were linked to fish advisories and included recommended actions. In short, culture was an important organizing strategy. Culture was also a key to the effective design of public engagement processes (e.g., locations of meetings, flavor of meeting, framing of discussions of decisions) (Silka, 2007; Silka 2002d).

Issues regarding solid waste management in Maine are equally important but involve different contexts and partners (e.g., town officials, state policy makers, community members, and industries that pay fees for getting rid of their trash, collecting trash, and recycling what otherwise might end up in landfills) (Blackmer et al., 2015; Isenhour et al., 2016; Isenhour et al., 2015). Towns are devoting ever larger portions of their budgets to waste management. These stakeholders rarely have the same views, the same needs, or access to the same information. Discussions are frequently contentious, in part because of previous conflicts. As a result, these participants are struggling to learn from each other across their differences and to devise strategies for dealing with a range of challenges. Environmental communication issues in partnership contexts such as these are strikingly different from those in new immigrant communities in urban areas of Massachusetts.

Partnerships come in so many forms that the variation can be dizzying as we attempt to make sense of and learn from them. Studies of researcher-stakeholder partnerships, for example, have shown the importance of attending to partnership preferences to help identify differences and find ways to tailor the participation process (Bieluch et al., 2016). Such efforts to compare partner preferences can provide data that inform the development of communication practices. The use of question sets, in conjunction with identified best practices for public participation, can also help adapt generalizable insights to a specific situation (McGreavy et al., 2013). For example, when Walker et al. (this volume) identify the theme in the forest plan revision that relationships and trust need focused attention, we can ask: How is trust built? What does trust look like for this group? In our work with shellfishermen and water quality regulators, we learned that trust is most effectively built out on the mudflat with those who dig clams and monitor water quality. Going out in the field helped demonstrate respect for the lived experiences of those stakeholders. The shared experiences helped build the relationships that enabled trust to form. Thus, pairing a higher-level recommendation with questions that help the recommendation be sensitive to the situation helps engage the tension between generalizability and context dependence.

Discussion, Implications, and Next Steps

Each of the highlighted areas—disciplines, topics, scales, decisions, and partners—demonstrates an important component of context dependence that has implications for building effective and generalizable approaches to environmental communication and public participation. This list is not exhaustive, however; nor do elements on the list always occur independently. At large scales, particular topics may be more likely to be the focus. Some topics may be more likely to be associated with specific kinds of partners. The knowledge needed to develop decision support tools will necessarily vary. And all of this is rife with complicating paradoxes. The need for solutions may sometimes be greatest at large scales (e.g., the need to reduce the global emission of greenhouse gases), but the local problems may be those for which actors experience the highest motivation and sense of control to solve. Or fear appeals—making climate change sound dire—may increase people's awareness but also lead them to shut down emotionally and avoid the topic (O'Neill & Nicholson-Cole, 2009). Issues of environmental communication too often have been seen as something that comes into play once the science has been completed and the results need to be translated (Burgess, Harison, & Filius, 1998). Such a view misses the fact that studies of communication processes need to be deeply integrated throughout the research process. The discipline of communication has a central role to play in moving environmental research and policy beyond its current state. In fact, the theories well understood by communication scholars are central to the generation of knowledge: using metaphors, analogies, and storytelling (Kahneman, 2011). Kahneman summarizes how the deployment of heuristics and rules of thumb tap into people's typical ways of processing information, including how the use of analogies helps people process shared information more readily.

Interdisciplinary uses of generative concepts, pursued in innovative ways, can move the field beyond approaches that have tended to be constraining and rate limiting (Larson, 2011). For example, approaches honed at the community level that assert that successful practices must be face-to-face to build trust are rate limiting because many types of environmental decisions do not occur at the community level. There is another way to learn, one that we have argued has greater promise because it provides a bridge across differences. As we have indicated, environmental and sustainability research is replete with overarching concepts that show great promise in stimulating cross-context learning, such as "wicked problems," "boundary spanning," "resilience," "honest broker," "the commons," "knowledge co-production," "citizen science," "stakeholder-researcher partnership," and "developing knowledge that is salient, credible, and legitimate." These concepts are not directives; nor are they roadmaps. Rather, they serve to stimulate reflection processes that can prompt innovation (Silka, 2014). One might say, "Here is a wicked problem in an urban setting and here is what we tried to do to strengthen environmental communication." Rather than applying the original approach unchanged to a rural setting, by using generative concepts, we are prompted to change the way we consider what constitutes a wicked environmental problem in a rural setting and what we need to do to create effective rural environmental communication.

Differences can seem insurmountable. When people look from the perspective of their rural setting at what was done in the urban setting, they might be tempted to say, "We can't do what they did. We don't have the same resources, and things are different here than in cities." People sometimes talk as if there are no lessons to be drawn on because the rural-urban differences are just too great. When we move the analysis to emphasize generative concepts, the conversation has the potential to change. Consider the generative concept of citizen science: it has been used to bring together scientists and community people who live near each other. Citizen science in an urban area may involve participants who can have considerable face time together. If we were to do citizen science in highly dispersed areas, it could still be done, but would likely need to take a different form. How might people think through what would work to achieve the overarching goals? Conversations could involve asking which aspects of citizen science are important: how people come together, who controls the coming together, how they communicate with each other, and so forth. Similarly, the overarching concept of "honest broker" (Pielke, 2007) could be used to move beyond a focus on rural-urban differences. Those in rural areas might say that there are certain kinds of contributors in urban areas who play a neutral role and help negotiate differences in environmental views, and we do not have these same kinds of contributors in our rural area. But bringing in the honest broker analysis might provide opportunities to innovate by thinking not about the exact type of contributor but about the overarching role. An honest broker in the urban context might look different from an honest broker in a rural context, but the concept can be a way to engage differences and innovate.

Overall, our point in this chapter is to find approaches to encourage learning about what works across highly diverse contexts. When moving across topics, what would the use of overarching concepts as bridging frameworks look like, and how can this be a way to foster the development of a corpus of useful environmental communication knowledge? The concept of citizen science was shaped by the challenge of involving citizens in collecting data that track shifting bird distributions. A very different topic is the emerging environmental problem of solid waste, with questions being raised about the rapid increase in trash production and how to get the message out about the need to reduce waste. This topic of trash has little in common with that of birds; thus, it can be hard to see how people studying birds can learn from people studying trash. But placing the focus on the overarching concept of citizen science can move beyond the specifics and help in envisioning how to involve people in assessing the magnitude of solid waste so as to lead to effective environmental communication. Rather than encouraging exactly the same strategy, the point here is to use the concept of citizen science to innovate around communication. Citizen science has worked for "sexy" topics like the study of rare birds, polar bears, and whales. Trash is far from being such a topic. Using the concept of citizen science can lead to new ways to thinking about waste but also new ways to imagine using citizen science in a broad range of unexpected contexts.

Finally, in this chapter we find that there is a place for research, but it may not be the role that is often envisioned where insights are identified and then applied across scale. One way the identification of generalizable insights can be tailored to diverse contexts is by looking for and using generative concepts from these research settings. Environmental communication research on public participation has opened up new understandings of what voice means in public participation settings and how, by paying attention to nuanced details, participation and inclusive decision making is enabled. Research thus could help to produce a set of generative concepts that can serve as prompts for the process of problem solving across complex public participation and collaboration contexts. Thus, our knowledge goal is best understood as that of developing a body of generative concepts that can then be actively "interrogated" to arrive at possible solutions to the challenges of learning across complex, different situations.

References

Ahmadi, B. V., Moran, D., Barnes, A. P., & Baret, P. V. (2015). Comparing decision-support systems in adopting sustainable intensification criteria. *Frontiers in Genetics, 6*(23), 1–15.

Allan, J. D., & Castillo, M. M. (2007). *Stream ecology: structure and function of running waters.* New York: Springer Science & Business Media.

Beierle, T., & Cayford, J. (2002). *Democracy in practice: Public participation in environmental decisions.* Washington, DC: Resources for the Future Press.

Berkes, F., & Ross, H. (2013). Community resilience: Toward an integrated approach. *Society & Natural Resources, 26*(1), 5–20.

Bieluch, K. H., Bell, K. P., Teisl, M. F., Lindenfeld, L. A., Leahy, J., & Silka, L. (2016). Transdisciplinary research partnerships in sustainability science: An examination of stakeholder participation preferences. *Sustainability Science*, 3, 1–18.

Blackmer, T., Criner, G., Hart, D., Isenhour, C., Peckenham, J., Rock, C., Rude, C., & Silka, L. (2015). *Materials and solid waste management in Maine: Past, present and future.* Senator George J. Mitchell Center Working Paper.

Boyle, K. J., Kuminoff, N. V., Zhang, C., Devanney, M., & Bell, K. P. (2010). Does a property-specific environmental health risk create a "neighborhood" housing price stigma? Arsenic in private well water. *Water Resources Research*, 46(3), 1–10.

Brown, R. R., Deletic, A., & Wong, T. H. F. (2015). Interdisciplinarity: How to catalyse collaboration. *Nature, 525*, 315–317.

Brown, V. A., Harris, J. A., & Russell, J. Y. (Eds.). (2010). *Tackling wicked problems: Through the transdisciplinary imagination.* London: Earthscan.

Brzustowski, T. (2012). *Why we need more innovation in Canada and what we must do to get it.* Ottawa: Invenire Books.

Burgess, J., Harrison, C. M., & Filius, P. (1998). Environmental communication and the cultural politics of environmental citizenship. *Environment and planning A, 30*(8), 1445–1460.

Burke, K. (1969). *A rhetoric of motives.* Berkeley, CA: University of California Press.

Calhoun, A. J., Jansujwicz, J. S., Bell, K. P., & Hunter, M. L. (2014). Improving management of small natural features on private lands by negotiating the science–policy boundary for Maine vernal pools. *Proceedings of the National Academy of Sciences, 111*(30), 11002–11006.

Calhoun, A. J. K., & Reilly, P. (2008). Conserving vernal pool habitat through community based conservation. In A. J. K. Calhoun and P. G. deMaynadier (Eds.), *Science and conservation of vernal pools in Northeastern North America* (pp. 319–344). Boca Raton, FL: CRC Press.

Cash, D. W., Clark, W. C., Alcock, F., Dickson, N. M., Eckley, N., Guston, D. H., . . . & Mitchell, R. B. (2003). Knowledge systems for sustainable development. *Proceedings of the National Academy of Sciences, 100*(14), 8086–8091.

Cash, D. W., Adger, W. N., Berkes, F., Garden, P., Lebel, L., Olsson, P., Pritchard, L. & Young, O. (2006). Scale and cross-scale dynamics: governance and information in a multilevel world. *Ecology and society, 11*(2), Article 8.

Chao, K., & Long, S. (2004). Youth participation in research: Weighing the benefits and challenges of partnership. *Race, Poverty, and the Environment, 11*(2), 43–45.

Clark, W. C., Tomich, T. P., van Noordwijk, M., Guston, D., Catacutan, D., Dickson, N. M., & McNie, E. (2011). Boundary work for sustainable development: Natural resource management at the Consultative Group on International Agricultural Research (CGIAR). *Proceedings of the National Academy of Sciences*, 200900231.

Clark, W. C., van Kerkhoff, L., Lebel, L., & Gallopin, G. (2016). *Crafting usable knowledge for sustainable development.* HKS Faculty Research Working Paper Series RWP16-005, Cambridge, MA: Harvard Kennedy School.

Coppens, N. M., Silka, L., Khakeo, R., & Benfey, J. (2000). Southeast Asians' understanding of environmental health issues. *Journal of Multicultural Nursing and Health, 6*(3), 31–38.

Cox, J. R., & Pezzullo, P. C. (2016). *Environmental communication and the public sphere* (4th ed.). Los Angeles, CA: Sage Publications.

Cox, R. & Depoe, S. (2015). Emergence and growth of the "field" of environmental communication. In A. Hansen & R. Cox (Eds.), *The Routledge handbook of environment and communication* (pp. 13–25). New York: Routledge.

Dahl, T. E. (1990). Wetland losses in the United States 1780s to 1980s. U.S. Department of the Interior, Fish and Wildlife Service, Washington, DC.

Daniels, S., & Walker, G. B. (2001). *Working through environmental conflict: The collaborative learning approach.* Westport, CT: Praeger Publishers.

Davenport, C., Gillis, J., Chan., & Eddy, M. (2015, December 12). Inside the Paris climate deal. *The New York Times.* Retrieved from http://www.nytimes.com/interactive/2015/12/12/world/paris-climate-change-deal-explainer.html?module=ConversationPieces&region=Body&action=click&pgtype=article&_r=0

Depoe, S. P., Delicath, J. W., & Elsenbeer, M. F. A. (Eds.). (2004). *Communication and public participation in environmental decision making.* Albany, NY: State University of New York Press.

Dietz, T., & Stern, P. C. (Eds.). (2008). *Public participation in environmental assessment and decision making.* Washington, DC: National Academies Press.

Druschke, C. G. (2013). Watershed as common-place: communicating for conservation at the watershed scale. *Environmental Communication: A Journal of Nature and Culture, 7*(1), 80–96.

Duncker, K. (1945). On problem solving. *Psychological Monographs, 58*, 1–113.

Easley, D., & Kleinberg, J. (2010). *Networks, crowds, and markets: Reasoning about a highly connected world.* Cambridge, UK: Cambridge University Press.

Eigenbrode, S. D., O'Rourke, M., Wulfhorst, J. D., Althoff, D. M., Goldberg, C. S., Merrill, K., . . . Bosque-Perez, N. A. (2007). Employing philosophical dialogue in collaborative science. *BioScience, 5*(1), 55–64. doi:10.1641/B570109

Endres, D. (2009). Science and public participation: An analysis of public scientific argument in the Yucca Mountain controversy. *Environmental Communication: A Journal of Nature and Culture, 3*(1), 49–75. doi:10.1080/17524030802704369

Endres, D., Sprain, L., & Peterson, T. R. (2008). The imperative of praxis-based environmental research: Suggestions from the Step It Up 2007 national research project. *Environmental Communication, 2*, 237–245.

Fox, J. L., & Cooper, C. (Eds.). (2013). *Boundary-spanning in organizations: Network, influence, and conflict.* Abingdon, UK: Taylor and Francis.

Gismondi, M., Connelly, S., Beckie, M., Markey, S., & Roseland, M. (Eds.). (2016). *Scaling up: The convergence of social economy and sustainability.* Edmonton, AB: AU Press.

Graybill, J. K., Dooling, S., Shandas, V., Withey, J., Greve, A., & Simon, G. (2006). A rough guide to interdisciplinarity: Graduate student perspectives. *BioScience, 56*(9), 757–764.

Grigg-Saito, D., Leong, S., Och, S., Silka, L., & Toof, R. (2008). Building on the strengths of a Cambodian refugee community through community-based outreach. *Health Promotion Practice, 9,* 415–425.

Guston, D. H. (2001). Boundary organizations in environmental policy and science: an introduction. *Science, Technology & Human Values, 26*(4), 399–408.

Hart, D. D., & Calhoun, A. J. (2010). Rethinking the role of ecological research in the sustainable management of freshwater ecosystems. *Freshwater Biology, 55*(s1), 258–269.

Hofstadter, D., & Sander, E. (2013). *Surfaces and essences: Analogy as the fuel and fire of thinking.* New York: Basic Books.

Isenhour, C., Wagner, T., Blackmer, T., Silka, L., Peckenham, J., Hart, D., & McRae, J. (2016). Moving up the waste hierarchy in Maine: Learning from "Best Practice" state-level policy for waste reduction and recovery. *Maine Policy Review, 24*(3), 15–29.

Isenhour, C., Blackmer, T., Hart, D., Silka, L., Peckenham, J., & Rudolph, J. (2015). *The future of materials management in Maine: Stakeholder engagement outcomes report.* Senator George J. Mitchell Center Working Paper.

Jansujwicz, J. S., Calhoun, A. J. K., & Lilieholm, R. J. (2013). The Maine vernal pool mapping and assessment program: engaging municipal officials and private landowners in community-based citizen science. *Environmental Management, 52,* 1369–1385.

Kahneman, D. (2011). *Thinking fast and slow.* New York: Farrar, Straus, and Giroux.

Kates, R. W., Clark, W. C., Corell, R., Hall, J. M., Jaeger, C. C., Lowe, I., McCarthy, J. J., Schellnhuber, H. J., Bolin, B. & Dickson, N. M. (2001). Sustainability science. *Science, 292*(5517), 641–642.

Larson, B. (2011). *Metaphors for environmental sustainability.* New Haven: Yale University Press.

Lansing, J. S. (2003). Complex adaptive systems. *Annual Review of Anthropology, 32,* 183–204.

Levesque, V. R., Calhoun, A. J. K, Bell, K. P., & Johnson, T. (2016). Turning contention into collaboration: engaging power, trust and learning in collaborative networks. *Society and Natural Resources, 30,* 245–260.

Levin, S. A. (1992). The problem of pattern and scale in ecology: the Robert H. MacArthur award lecture. *Ecology, 73*(6), 1943–1967.

Malone, P. M. (2009). *Waterpower in Lowell: Engineering and industry in nine-teenth-century America.* Baltimore: Johns Hopkins University Press.

Marion, P. (2014). *Mill power: The origin and importance of Lowell National Historical Park.* Lanham, MD: Rowan and Littlefield.

Marrone, J. A. (2010). Team boundary spanning: A multilevel review of past research and proposals for the future. *Journal of Management, 26*(4), 911–940.

McCaffrey, T. (2012). Innovation relies on the obscure: A key to overcoming the classic functional fixedness problem. *Psychological Science, 23*(3), 215–218.

McGreavy, B., & Hart, D. Sustainability science and climate change communication (2017, May). In M. Nisbet (Ed.), *Oxford encyclopedia of climate change communication.* New York: Oxford University Press. doi:10.1093/acrefore/9780190228620.013.563

McGreavy, B., Druschke, C.G., Sprain, L., Thompson, J., & Lindenfeld, L. (2017). Praxis-based environmental communication training: innovative activities for problem solving. In T. Milstein, M. Pileggi, & E. Morgan (Eds.), *Environmental communication pedagogy and practice* (pp. 229–38). London: Routledge.

McGreavy, B., Calhoun, A. J. K., Jansujwicz, J., & Levesque, V. (2016). Citizen science and natural resource governance: Program design for vernal pool policy innovation. *Ecology and Society,* 21, Article 48.

McGreavy, B., Lindenfeld, L., Bieluch, K., Silka, L., Leahy, J., & Zoellick, B. (2015). Communication and sustainability science teams as complex systems. *Ecology and Society, 20*(1), Article 2. doi:10.5751/ES-06644-200102

McGreavy, B., Hutchins, K., Smith, H., Lindenfeld, L., & Silka, L. (2013). Addressing the complexities of boundary work in sustainability science through communication. *Sustainability, 5*(10), 4195–4221. doi:10.3390/su5104195

McNie, E. C., Parris, A., & Sarewitz, D. (2016). Improving the public value of science: A typology to inform discussion, design and implementation of research. *Research Policy, 45*(4), 884–895.

Meyer, S. R., Johnson, M. L., Lilieholm, R. J., & Cronan, C. S. (2014). Development of a stakeholder-driven spatial modeling framework for strategic landscape planning using Bayesian networks across two urban-rural gradients in Maine, USA. *Ecological Modelling, 291,* 42–57.

Messer, K. D., Schulze, W. D., Hackett, K. F., Cameron, T. A., & McClelland, G. H. (2006). Can stigma explain large property value losses? The psychology and economics of Superfund. *Environmental and Resource Economics, 33*(3), 299–324.

Miller, T. R., Baird T. D., Littlefield, C. M., Kofinas, G., Chapin, F. S., & Redman, C. L. (2008). Epistemological pluralism: reorganizing interdisciplinary research. *Ecology and Society, 13*(2). Retrieved from http://www.ecologyandsociety.org/vol13/iss2/art46/

Moser, S., & Dilling, L. (Eds.). (2007). *Creating a climate for change: Communicating climate change and facilitating social change.* New York: Cambridge University Press.

Newell, W. H. (Ed.). (1998). *Interdisciplinarity: Essays from the literature.* New York: The College Board.

Norberg, J., & Cumming, G. S. (Eds.). (2008). *Complexity theory for a sustainable future.* New York: Columbia University Press.

O'Neill, S., & Nicholson-Cole, S. (2009). "Fear won't do it": Promoting positive engagement with climate change through visual and iconic representations. *Science Communication, 30*(3), 355–379.

O'Rourke, M., Crowley, S., Eigenbrode, S. D., & Wulfhorst, J. D. (2013). *Enhancing communication and collaboration in interdisciplinary research.* Thousand Oaks, CA: Sage Publications.

Ostrom, E. (2007). A diagnostic approach for going beyond panaceas. *Proceedings of the National Academy of Sciences, 104*(39), 15181–15187. doi:10.1073/pnas.0702288104

Owen, D., & Apse, C. (2014). Trading dams. *UC Davis Law Review, 48,* 1043.

Palmer, M. A. (2012). Socioenvironmental sustainability and actionable science. *BioScience, 62,* 5–6.

Pezzullo, P. C. (2007). *Toxic tourism: Rhetorics of pollution, travel, and environmental justice.* Tuscaloosa: University of Alabama Press.

Pielke, R., Jr. (2007). *The honest broker: Making sense of science in policy and politics.* Cambridge, UK: Cambridge University Press.

Pretty, J. (2011). Interdisciplinary progress in approaches to address social-ecological and ecocultural systems. *Environmental Conservation, 38*(2), 127–139. doi:10.1017/S0376892910000937

Pho, T., Gerson, J., & Cowan, S. (Eds.). *Southeast Asian refugees in the mill city: Changing families, communities, institutions thirty years afterward.* Lebanon, NH: University Press of New England.

Sandler, R. D., & Pezzullo, P. C. (Eds.). (2007). *Environmental justice and environmentalism: The social justice challenge to the environmental movement.* Cambridge, MA: MIT Press.

Senecah, S. L. (2004). The trinity of voice: The role of practical theory in planning and evaluating the effectiveness of environmental participatory processes. In S. P. Depoe, J. W. Delicath, & M. F. A. Elsenbeer (Eds.), *Communication and public participation in environmental decision making* (pp. 13–33). Albany, NY: State University of New York Press.

Silka, L. (2002a). Combining history and culture to reach environmental goals. *New Village: Building Sustainable Cultures, 3,* 4–11.

Silka, L. (2002b). Environmental communication in refugee and immigrant communities: The Lowell experience. *Applied Environmental Education and Communication, 2,* 105–112.

Silka, L. (2002c). Immigrants, sustainability and emerging roles for universities, *Development Journal of the Society of International Development, 45*(3), 119–123.

Silka, L. (2002d). A university enters into its regional economy: Models for integrated action with refugee and immigrant communities. In J. L. Pyle & R. Forrant

(Eds.), *Globalization, universities, and issues of sustainable human development* (pp. 154–72). Cheltenham, UK: Edgar Elgar Press.

Silka, L. (2007). Immigrants in the community: New opportunities, new struggles. *Analyses of Social Issues and Public Policy, 7*(1), 1–17.

Silka, L. (2014). "Silos" in the democratization of science. *Journal of Deliberative Mechanisms in Science, 2*(1), 1–14.

Silka, L., Kelly, R., & Ward, J. S. (Eds.). (2014). Innovation: Special Issue of *Maine Policy Review, Maine Policy Review, 23*(1), 1–94.

Smith, H. M., & Norton, T. (2013). "That's why I call it a task farce": Organizations and participation in the Colorado Roadless Rule. *Environmental Communication: A Journal of Nature and Culture, 7*(4), 456–474.

Sowards, S. K. (2012). Environmental justice in international contexts: Understanding intersections for social justice in the twenty-first century. *Environmental Communication: A Journal of Nature and Culture, 6*(3), 285–289.

Sprain, L., & Timpson, W. (2012). Pedagogy for sustainability science: Case-based approaches for interdisciplinary instruction. *Environmental Communication: A Journal of Nature and Culture, 6*, 532–550.

Steinberg, T. (1991). *Nature Incorporated: Industrialization and the waterways of New England.* Cambridge, UK: Cambridge University Press.

Stuart, S. N., Chanson, J. S., Cox, N. A., Young, B. E., Rodrigues, A. S., Fischman, D. L., & Waller, R. W. (2004). Status and trends of amphibian declines and extinctions worldwide. *Science, 306*(5702), 1783–1786.

Van de Ven, A., Polley, D., Garud, R., & Venkataraman, S. (1999). *The innovation journey.* New York: Oxford University Press.

Walker, G. B., Senecah, S. L., & Daniels, S. E. (2006). From the forest to the river: citizens' views of stakeholder engagement. *Human Ecology Review, 13*, 193–202.

Waring, T. (2012). Wicked tools: The value of scientific models for solving Maine's wicked problems. *Maine Policy Review, 21*(1), 30–39.

Wolf, J., & Moser, S. C. (2011). Individual understandings, perceptions, and engagement with climate change: Insights from in-depth case studies across the world. *WIREs Climate Change, 2*, 547–569. doi:10.1002/wcc

Section I

Exploring Dimensions of Participation Within Policy Frameworks

Chapter 2

Listening and Learning

Stakeholder Views of Participation and Communication in Forest Planning

GREGG B. WALKER, STEVE DANIELS, SHARON TIMKO, CARMINE LOCKWOOD, AND SUSAN HANSEN

In their best-selling negotiation text, *Getting to Yes*, Roger Fisher, William Ury, and Bruce Patton assert that "without communication there is no negotiation" (2011, p. 32). Similarly, in public policy decision situations, one could claim that "without communication there is no public participation." Public participation is fundamentally about communication—citizens expressing their view about potential policy alternatives. In environment and natural resource management decision situations, communication activity is an essential feature of public participation (Cox, 2012). Citing a number of scholars (Cox, 2009; Beierle & Crayford, 2002; Dietz & Stern, 2008), Tracylee Clarke and Tarla Rai Peterson, in *Environmental Conflict Management* (2015), assert that public participation provides opportunities for citizens and groups to influence decisions about the environment. They explain that public participation includes "any organized process or mechanism, intentionally instituted, organized, or adopted by government agencies to engage the public in administrative decision making, environmental assessment, planning, management, monitoring, and evaluation" (p. 91).

As part of the public participation strategies that agencies and organizations employ, communication occurs in numerous forms such as media (e.g., television, radio), technology (e.g., cell phones), social media (e.g.,

Facebook, Twitter, blogs), and face-to-face (e.g., public hearings, workshops, dialogues, appeals) (Burgess & Burgess, 1997; Senecah, 2004; Phillips et al., 2012; Clarke & Peterson, 2016). Some of these activities, such as public hearings and comment letters, represent a conventional, "old school" paradigm, while others are innovative "new school" methods that may emphasize face to face interaction (e.g., charrettes, workshops, world cafes), technology (e.g., collaborative mapping), or social media (Walker et al., 2015; Clarke & Peterson 2016).

This chapter relates to both conventional and creative approaches to public participation and communication. It does so by reporting on the views of people who care about environmental and natural resource management decisions and want to be involved specifically in forest management planning and activities. The discussion begins by addressing the idea of public participation and its significance in developing environmental policy. In doing so, it highlights the significance of the United States' National Environmental Policy Act as a model for public participation policies throughout the world and as a precursor for context-specific legislation and policies. The commentary highlights public participation efforts in a variety of countries, contending that public participation is a global concept (if not a global practice).

The chapter subsequently transitions to presenting citizen views of public participation and communication, drawn from comments recorded at "listening sessions" conducted as part of Forest Plan revision efforts on two National Forests in the United States. Just as Speaker of the House Tip O'Neill remarked that "all politics is local," the most meaningful public participation often occurs locally as well. Although this chapter focuses on communities connected to two U.S. National Forests, one can imagine the importance of local voices wherever public participation occurs. From an analysis of the comments of National Forest listening session participants, nine themes emerge that can guide the development and implementation of participation and communication strategies and activities, perhaps both domestically and internationally.

Public Participation

In the United States, public participation in environmental and natural resource management was relatively limited prior to 1970. Federal government agencies followed a "managerial model" in which government officials "were entrusted to identify and pursue the common good" (Beierle & Cayford,

2002, p. 2). Government agencies did not have to ask citizens for ideas; they knew what was best for them.

The Administrative Procedures Act (APA) of 1946 included the idea of public participation as part of federal policy decision making. Beierle and Cayford have explained that the APA required that agencies, when creating law through rule making, "provide public notice about the rules they are proposing, information on which the rules are based, an opportunity for public comment on those rules, and judicial review of the rulemaking process" (2002, p. 3). The APA began a change process and provided a foundation for future action. In the 1960s and 1970s, through the Johnson and Nixon presidential administrations, other public participation legislation was enacted, including the Freedom of Information Act in 1966 and the Federal Advisory Committee Act of 1972 (Long & Beierle, 1999; Carey, 2013).

In terms of public participation in environmental and natural resource management decisions, no legislation was more significant during this period than the National Environmental Policy Act, or NEPA. In the late 1960s, "Congress faced a need not only to respond to the values underlying the growing concern over a deteriorating environment but as much as possible to recognize and harmonize the diversity of values and concepts present in American society that related to the environment" (Caldwell, 1999, p. 2). NEPA drew strength from public attitudes and values about nature and the environment, influenced in part by the new environmentalism movement and books like Rachel Carson's *Silent Spring* (Lindstrom & Smith, 2001; Daley, 2012). As Kalen notes, "Throughout the 1950s and 1960s, our society, including Congress, became acutely aware of the growing environmental crisis" (2009, p. 492).

NEPA's design responded to public values and interests in a ground-breaking way; it incorporated participation into the environmental planning process. The act required environmental decision makers, planners, and administrators to reach out to civil society and provide opportunities for the public to be involved (Walker et al., 2015). According to the Natural Resources Defense Council, "NEPA is democratic at its core. In many cases, NEPA gives citizens their only opportunity to voice concerns about a project's impact on their community. . . . NEPA is designed to ensure that the public has informed access and input into federal agency decisions that could affect the human or natural environment" (Natural Resources Defense Council, cited in Walker et al., 2015).

The NEPA process requires agencies to develop project or management alternatives and provide opportunities for citizens to review and comment on alternatives. Some agencies have welcomed citizen-generated alternatives as part of their NEPA process. For example, when two of the authors (Daniels and Walker) worked with the Chugach National Forest on its Forest Plan revision process, the Forest Plan revision team encouraged members of the community to develop their own alternative as well as comment on the alternatives the Forest team prepared (Daniels & Walker, 2001).

As Walker, Daniels, and Emborg explain in a recent essay, "NEPA elevated the importance of public participation in environmental planning and decision-making, particularly through Council of Environmental Quality or CEQ (created as a part of NEPA) policies" (2015, p. 131). The CEQ has worked with federal agencies to provide guidance in the application of NEPA in planning and project work. For example, in 2007, the CEQ published *Collaboration in NEPA: A Handbook for NEPA Practitioners* to assist people within and outside federal agencies to develop and apply collaborative processes consistent with and as a part of NEPA work. As the introduction of the handbook explains:

> One of the primary goals of the National Environmental Policy Act (NEPA) is to encourage meaningful public input and involvement in the process of evaluating the environmental impacts of proposed federal actions. This once innovative feature of the 1970 landmark legislation has become routine practice for some NEPA review processes. However, the full potential for more actively identifying and engaging other Federal, Tribal, State and local agencies, affected and interested parties, and the public at large in collaborative environmental analysis and federal decision-making is rarely realized. (Council on Environmental Quality, 2007, p. 1)

According to NEPA scholar Lynton Caldwell, prior to NEPA, no nation had passed comprehensive environmental policy legislation (Caldwell, 1999). Although some countries (particularly developed nations) were beginning to address environmental issues in the two decades after the Second World War, NEPA had no equivalent in 1969, nor was there a model or precedent for this action (Caldwell, 1999). NEPA has since become a model both internationally and nationally, with nations and U.S. states establishing environmental planning policies patterned after NEPA (Walker et al., 2015). In addition to public participation requirements, other NEPA innovations that have been copied include planning and project distinctions (e.g., when to conduct a full environmental analysis) and consulting with other agencies and indigenous communities (e.g., Tribes).

The NEPA process and the CEQ policies emphasize the importance of providing citizens with opportunities to influence environmental policy decisions, typically with some combination of written comments and meetings. Other countries, both developing and developed, have included public participation in their policy decision-making processes, such as through sustainable development and environmental management projects that address governance concerns. For example, as part of its governance strategy, the World Bank Institute promotes "multi-stakeholder collaborative action," recognizing that "successful development requires more than technical solutions. It requires getting individuals, groups, and organizations to work together to achieve a complex set of objectives" (World Bank Institute, 2014).

In the United Kingdom, for example, participation has been a component of Landscape Character Assessment (LCA), an approach for "devising indicators to gauge landscape change and to inform regional planning, local development, environmental assessment and the management of protected landscapes" (Natural England, 2014; Tudor, 2014). In its earliest applications, LCA was regarded by planners as a process that emphasized work by professionals. As LCA efforts evolved, planners recognized the need for citizen and stakeholder involvement (Swanwick, 2004). LCA work typically included stakeholder engagement in a participatory planning process (Swanwick, 2004), although not at the community level (Warburton, 2004). Nongovernmental organizations have provided stakeholder input, but opportunities for the general public have not been mainstream in LCAs and similar efforts (Butler & Berglund, 2014; Kamphorst et al., 2017).

In Denmark, LCA is part of national parks development. The Danish government has adapted the LCA process to serve parks development, with the intention of involving stakeholders. But as of 2009, "the part of the Danish LCA that relates to participatory planning has not yet been developed, despite the recognition that it may provide important contributions to the assessment of local landscape character" (Casperson, 2009, p. 34). Since that time, Danish organizations (governmental and nongovernmental) have invested more resources in public participation in environmental planning, exemplified by the work of Danish Water Councils (Graversgaard et al., 2016).

In 2003, China established the Environmental Impact Assessment (EIA) Law. It includes provisions for public participation. China's EIA requires public notice of environmental assessments, document disclosure, public comment periods, the possibility of public hearings or meetings, and responses to submitted comments (Schulte, 2013). The promise of public participation, though, as outlined in the EIA differs from practice, perhaps because public participation can imply democratic governance. "Mass environmental protests in China have increased at a rate of around 29% per year since 1996, and by a staggering 120% in 2012," the China Dialogue website reports, adding that "a common complaint is that the legal requirements for transparency and public participation in environmental decision-making are often ignored" (Schulte, 2013).

Since the 1992 Earth Summit (the United Nations Conference on Environment and Development) in Rio de Janeiro, Brazil, nations throughout Sub-Saharan Africa have implemented environmental impact assessment (EIA) policies. These EIA programs typically include some provision for public participation, but a "common weakness in many EIAs is the continued lack of public participation and involvement in the process" (Kakonge, 2006, p. 20), primarily due to corruption and a lack of capacity (Kakonge, 2013). In a study of public involvement in Kenya EIA efforts, Okello and colleagues have reported that "interview results indicate a diverse list of constraints such as poor information sharing, lack of consultation, incomprehensible language, lack of familiarity with EIA guidelines, and lack of institutional and regulatory capacity hinder serious public involvement" (2008, p. 1; see Marara et al., 2011 and Utembe, 2015 for case studies). Correspondingly, "in most African States, there are still challenges related to relaying EIA results to the stakeholders, communities and decision makers" (Kakonge, 2013). There are exceptions, such as Botswana, which has consulted with key stakeholders during an EIA process, even though the EIA process in Botswana is ad hoc rather than formalized (Kakonge, 2006).

Regional initiatives have also promoted public participation. The United Nations Economic Commission for Europe (UNECE) Convention on Access to Information, Public Participation in Decision-Making and Access to Justice in Environmental Matters was adopted in June 1998 in the Danish city of Aarhus at the Fourth Ministerial Conference as part of the "Environment for Europe" process. It entered into force at the end of October 2001. Known as the "Aarhus Convention," the agreement establishes a number of rights of the public (individuals and their associations) with regard to the environment. These include (1) the right of everyone to receive environmental information that is held by public authorities and (2) the right to participate in environmental decision making (European Commission, 2016). In 2003, two Directives concerning these two "pillars" of the Aarhus Convention were adopted; they were to be implemented in the national law of the European Union (EU) member states in June 2005 (European Commission, 2016). EU countries are expected to provide information and participation opportunities as part of environmental planning.

Just as European Union members are obliged to invest in public participation, so, too, are members of the Organization of American States (OAS). OAS member countries are expected to adhere to the following mandates pertaining to governance and participation. Mandate 79, adopted as part of the Port of Spain Declaration in 2009, states that OAS countries "recognize the role of governance at the local level as a tool for strengthening democracy and sustainable development [and] affirm the importance of enhancing decentralisation, local government, and citizen participation" (Summit of the Americas Secretariat, 2005).

This mandate appears in the democracy and governance categories of OAS documents, accompanied by Mandate 36, endorsed in the 2005 Mar del Plata Declaration. The Mandate states, "We consider it essential to strengthen broad, transparent, and inclusive social dialogue with all concerned sectors of society, at the local, national, regional, and hemispheric levels . . . an important and basic instrument to promote and consolidate democracy and to build societies with inclusion and social justice" (Summit of the Americas Secretariat, 2005). Subsequent to the declaration, OAS member nations have acted on the Mar del Plata Declaration, involving civil society in a number of policy areas, including the environment and development (Joint Summit Working Group, 2009).

In the United States, where various public participation policies have been operating for decades at the federal, state, and local levels, the record is uneven. Although public participation strategies in environmental and natural resource policy situations are intended to improve decision making and strengthen legitimacy, they often fall short of achieving these objectives (Webler et al., 2001; Dietz & Stern 2008). Public participation is just one factor in complex planning and decision processes, with stakeholder expectations, decision space, and regulatory and technical demands seemingly at odds, a part of what Daniels and Walker (2001) have discussed as the "paradox of public involvement" (Walker et al., 2015).

This "paradox of public involvement" seems inherent in complex and controversial policy decision situations. The paradox stems from two expectations that both policy makers and citizens hold: (1) policy decisions are based on the best available scientific and technical information; and (2) people who have a "stake" in the decision situation deserve opportunities to participate in the development of the decision and contribute local and indigenous knowledge. As Daniels and Walker explain, "the juxtaposition between technical competence and open process is a defining characteristic of American policy formation . . . These dual goals—technical competence and participatory process—create a compelling dynamic between a narrow politics of expertise and a broad politics of inclusion, a dynamic that cuts across public policy disputes such as nuclear waste disposal, health care, and land management" (2001, p. 4).

Another way to consider the paradox features scale. The emphasis on scientific and technical information and competence typically relies on "scale down" (or "top down") approaches, with planning centralized in government ministries or agencies and related institutions (such as universities). The "technical experts" determine what is best for the public. In contrast, public participation implies "scaling up" (or "bottom up"), with citizens and communities attempting to communicate their knowledge, values, interests, concerns, and even decision alternatives. Addressing scale, though, is not an "either-or" decision. Rather, environmental governance and decision making involve actions at multiple scales. Participation of the public—of villages and communities—provides a foundational scale at the grassroots level (Armitage et al., 2007; Young, 2012; Nagendra & Ostrom, 2012; Anderson, 2013).

"Scaling up" has emerged as an important component for achieving sustainable development goals (SDGs). Since the 1980s, development theorists and practitioners have increasingly emphasized the importance of scaling up development programs, relying on local formal and informal institutions and networks for both design and implementation (Nelson et al., 2009). Scale (and scaling up) has also gained attention in the climate change policy arena, particularly in areas such as community forestry (Garcia-Lopez, 2013) and adaptation in developing countries (Schipper et al., 2014).

Since the passage of NEPA, government agencies in the United States have revised, refined, and expanded their approaches to planning and decision making with guidelines that implicitly recognize scale, using approaches that include both the best available science and public participation. Since the early 1990s, public land management agencies such as the USDA Forest Service and the USDI Bureau of Land Management, for example, have issued guidelines from their Washington, DC, office to field staff, recommending or requiring collaboration with stakeholders, Tribes, and other agencies (USDA-Forest Service, 1997 & 2000; USDI-Bureau of Land Management, 2003).

Within the Forest Service, one such directive comes in the form of a "planning rule," procedural requirements for revising forest management plans.

This planning rule typically calls for community-based or grassroots public participation and emphasizes communication and collaborative engagement with citizens and stakeholders. As such, it encourages "scaling up" as part of the Forest Planning process. Such "scaling up" can be fostered, in part, by talking with and listening to citizens and stakeholders for their ideas about participation, collaboration, and communication. The remainder of this chapter presents on a project that asked citizens and members of sovereign Indian Tribes for their views. Although only one project, the themes that emerged from the conversations may be applicable to any situation where grassroots stakeholder participation is essential.

Public Participation, and the New Planning Rule, and Two USA National Forests

This project began in early 2012, when the USDA Forest Service released a new Planning Rule designed to guide National Forests as they revised their Forest Plans. One of the eight purposes and needs of the 2012 Planning Rule is to "provide for a transparent, collaborative process that allows effective public participation" (36 CFR Part 219). In 2013, two National Forests in a western region of the United States informally initiated the process of Forest Plan revision, in accordance with the new Planning Rule. The Forests invited a consulting team from the National Collaboration Cadre of the USDA-Forest Service to help them initiate their public participation process.

In the United States, each National Forest is governed by a Forest Land and Resource Management Plan (Forest Plan) as specified in the 1976 National Forest Management Act (NFMA). Forest Plans set desired conditions, standards, and guidelines for the management, protection, and use of the Forest. National Forests throughout the United States have been operating with Forest Plans adopted in the in the late 1980s and early 1990s. These Forest Plans should be updated periodically and revised every 10 to 20 years. The Forest Plan revision process responds to changed conditions on the specific National Forest, the best available science, public values, legal requirements, and new and emerging topics on that Forest.

The USDA-Forest Service National Collaboration Cadre (Collaboration Cadre), established in 2008, provides coaching and training assistance to National Forests, their communities, and interested stakeholders to help them engage in meaningful public participation and effective collaboration. The Collaboration Cadre is a network of consultants and Forest Service staff— people from around the United States who represent different perspectives regarding collaborative processes: agency, community, and academic. Consequently, the Collaboration Cadre offers a unique approach that models and blends the perspectives of the Forest Service, communities, and universities through the guidance of team members with extensive experience in collaborative public participation and outreach.

Since its inception, Collaboration Cadre teams have worked with National Forests in every region of the United States. Under the 2012 Planning Rule, a number of National Forests have invited Collaboration Cadre teams to assist them with public participation and communication activities as a foundational step in Forest Plan revision (e.g., the Chugach National Forest in Alaska; the El Yunque National Forest in Puerto Rico, and the Tonto National Forest in Arizona). The forests have requested varying levels of support from the Collaboration Cadre ranging from assessment, coaching, and training to designing and facilitating their initial public engagement events (e.g., community workshops).

To support and assist National Forest staff and Forest communities, Collaboration Cadre teams typically employ a three-pronged approach. First, the Cadre team offers a training workshop for Forest Leadership and key staff. Second, the Cadre team conducts assessment activities—such as listening sessions or community conversations—with Forest employees, key community stakeholders, Tribes, and relevant government organizations. Through the listening sessions and conversations, Cadre team members learn from participants about the forest management situation and can assess the collaborative potential of Forest Plan revision or a specific project, including opportunities and constraints (Daniels & Walker, 2001). Although NEPA does not require collaboration, the Planning Rule states its importance. Third, the Cadre team members design and facilitate community events, such as workshops or dialogues. Event participants include a diversity of interested parties (government and nongovernment), as well as National Forest staff.

A community workshop or dialogue session may be planning, project, or landscape focused. In the case of a Forest Plan revision process, workshops emphasize parties working together to develop a public participation, communication, and outreach strategy for revising the Forest Plan. Following public participation workshops, National Forest staff continue to develop and implement participation activities, typically assisted by a professional facilitator. The Forest Plan revision process is designed to take three or four years, with public participation, communication, and outreach activities occurring throughout.

The Forest Leadership Team (FLT) and planning staff of the two Forests made a commitment to engaging Tribes, stakeholders, communities, and government organizations through the Forest Plan revision process—early and at appropriate times during the 2014–2018 Plan revision period. At the outset of the Plan revision effort, the two Forests invited a Collaboration Cadre team to provide a staff training, conduct listening sessions, and design and facilitate community workshops. Both the listening sessions and workshops occurred in various communities throughout the Forests' regions.

Listening Sessions—Purposes and Procedures

Based on successful experiences with numerous collaborative planning efforts, and with the endorsement of National Forest leadership, Cadre members use listening sessions as an important first step in understanding the nature of forest management situations. The sessions offer Tribes, community stakeholders, organizations, and Forest Service staff an opportunity to 1) discuss how communities and stakeholders work together and with the Forest; 2) describe opportunities for and constraints on collaborative engagement; 3) exchange ideas about how to conduct public involvement and communication; 4) reveal topics or trends that might emerge during plan revision; and 5) and begin networking, sharing knowledge and experience, and identifying participants for future collaborative work.

Listening sessions for the two National Forests took place in 2014, about five weeks before the community workshops. Sessions ranged between 60 to 90 minutes. A total of 270 people participated in 32 listening sessions at 16 sites in communities within and near the National Forests (including nearby urban areas). Listening sessions were held at a variety of facilities, including fire stations, libraries, community centers, and Forest Service District offices.

Tribal members participated in one or more of four Tribal sessions held at various sites selected by the Tribes. Five sessions were internal conversations with National Forest personnel. Approximately 45 employees attended the internal listening sessions. Participating employees were a mix of resource professionals representing most of the major programs (e.g., fire, recreation, soils, habitat, minerals, plants) of the two Forests.

Twenty-three listening sessions, all at different locations, were attended by stakeholders and community members only. These attendees represented a wide range of interests, experiences, and backgrounds, including ranchers, summer and winter recreationists, environmental activists, local government officials, business owners, and retirees. In addition to these meetings, Cadre team members talked by phone with a number of people who were not able to attend a listening session.

Listening session participants were self-selected; they chose to attend. Many participants responded to an invitation from one or both of the National Forests. Others learned about the listening sessions from another person or an organization. The listening sessions included a large number of people who had not signed up; participation exceeded expectations. Everyone who wanted to participate was welcomed; no one was turned away. For example, three people had signed up for a Saturday morning listening session at a library. Sixteen people showed up, including secondary school students.

Two Cadre team pairs (authors Walker, Timko, Lockwood, and Hansen) were designated to facilitate each listening session, although, because of a medical emergency, eight meetings were facilitated by a single Cadre member. The format of the sessions was informal and conversational, guided by broad discussion themes. The facilitators began by briefly describing Forest Planning and the two western National Forests' intent to engage interested stakeholders, community members, and Tribes at feasible and appropriate times throughout the plan revision process. Cadre members presented the listening sessions as "Meeting Zero" in the four-year plan revision process.

After welcoming the participants and presenting a brief explanation of the Forest Plan Revision (FPR) process, Cadre members asked, "Have you been involved a prior forest plan revision process?" Most had not. The Cadre team then guided a conversation that addressed four questions/topics:

1. **Relationship** with the National Forest (connection to; experience with)

2. **Communication** and the National Forest (communication activities, appropriateness and effectiveness, experiences . . . what does the Forest do well; how can it improve)

3. **Participation** and the National Forest (public participation activities, appropriateness and effectiveness, experiences . . . what does the Forest do well; how can it improve)

4. **Issues** that may emerge as part of FPR

This general format was modified and adapted for particular Listening Session groups, such as the Tribes and the Internal Staff.

Cadre members took detailed notes during the listening sessions, recording the comments of the public to the best of their ability. The listening session comments and findings reflected the views and opinions of those who attended and not necessarily of the larger community. Cadre teams did not verify the accuracy of the participant statements or challenge participants on the validity of their comments. Regardless, participant contributions had value. When presenting their listening session findings to Forest staff, Cadre team members noted that the accuracy of participant statements was less important than what could be learned and understood about the perceptions, information, and judgments people had shared among themselves and with other parties.

While meaningful, the listening session comments provide part but not all of the story of forest management. Forest management situations are inherently complex (Walker et al., 2006; Daniels & Walker, 2012). Many listening session participants are, understandably, focused primarily on one or some of the forest management topics a Forest Plan covers. Few of the listening session attendees have prior experience with or knowledge of a Forest Plan revision process. Still, the listening sessions provide a constructive opportunity for Stakeholders, Forest staff, and Tribal members to communicate their interests and concerns and to listen and learn from one another.

The listening sessions collectively produced two "deliverables": 1) a four-page briefing paper that was written and presented to Forest Staff a day after the last listening sessions, and 2) a listening sessions report. The briefing paper provided a foundation for that report. With the briefing paper as a starting point, one member of the Cadre team (Walker) collected all the listening session notes, examined them in detail, and subsequently prepared a preliminary analysis. That analysis, along with Cadre team members' notes, was reviewed by the three other members of the Cadre team (Timko, Lockwood, and Hensen). Collectively, the Cadre team generated and revised a number of themes and observations that emerged from participants' comments in the listening sessions. The following section is organized around these themes. Some of themes have been presented to the Leadership and Planning teams of the two Forests in the listening sessions report.

Themes and Observations That Emerged from the Listening Sessions

When viewed collectively as a set, the themes make the case for collaborative work with parties (Tribes, stakeholders, communities, organizations) as a central feature of Forest Plan revision. There is collaborative potential in the Forest Plan revision process, but roles need to be clarified, expectations managed, messages communicated consistently, and decision space clarified. Decision space clarifies what stakeholders can contribute. What is open to negotiation and influence? What aspects of the situation are "on the table" for discussion, and what matters are not? For example, one of Daniels and Walker's earliest field projects involved a national recreation area managed by the U.S. Forest Service. As they designed and facilitated community workshops about this recreation area, they asked Forest Service officials to clarify what was "within" and "outside" the decision space. Recreation area curfews were within the decision space, while threatened and endangered species were not (Daniels & Walker, 1996, reported in Walker et al., 2015). Furthermore, National Forest staff needed to demonstrate that they have listened to stakeholders, accounted for their ideas, and followed through on commitments that Forest staff have made.

As the Cadre team members constructed, reviewed, and integrated the listening session notes from the Community and Tribal meetings, a number of themes and observations emerged. Some themes and observations were drawn from the entire set of listening sessions; others came from listening sessions with the Tribes. The themes and observations reflect the Cadre Team members' recording of participants' comments and the interpretation of those comments. Although some of these themes may seem obvious to readers familiar with public participation and stakeholder engagement literature, these themes were important for National Forest staff to acknowledge.

1. Many people want to work with the National Forests. Throughout the listening sessions, participants expressed a willingness to work with the National Forests on planning efforts that were meaningful to them. People were motivated to participate and be involved. The number of listening session participants exceeded expectations, in part, because people cared about the National Forests. These places were important to them: some sites were sacred places; others were significant to generations of families; still others contributed to people's livelihoods. Even those participants who were skeptical and frustrated in light of past planning and project work talked about mobilizing volunteers, sharing information, and assisting the Forests where appropriate.

Sample statements illustrating this theme included:

- An Indian Tribe has good working with relationships with District Ranger staff—especially the District Ranger.

- A Forest's CFLRP (Collaborative Forest Landscape Restoration Project) is a good example of people working together—it has a good website and good advanced planning.

- A local horse club has an MOU (Memorandum of Understanding) with the Forest; every year they get together with the Forest to develop a plan to help with erosion issues on trails.

2. Many parties seem to have a single topic or limited topic focus. Participants in the listening sessions often focused on one topic or a limited number of items. They were concerned primarily with substantive topics and did not want to discuss process at the expense of their topic. Ranchers wanted to comment on grazing; motorized recreation enthusiasts wanted to talk about roads; business owners and local officials wanted to discuss economic development. Although participants may have been familiar with the Forest Service's multiple-use mandate, many did not seem aware of the complexity of forest management. For example, most of the participants in one listening session were interested primarily in off-road motorized vehicle access. At subsequent community workshops, they were introduced to the interests of other stakeholders and the need to respect a range of views.

Sample statements illustrating this theme included:

- To have meaningful meetings, the Forest Service needs to recognize the rights of those who predate the Forest Service. Acknowledge the rights of ranchers.

- There is a public perception that the Forests deem it easier to close roads/trails/campgrounds than to maintain them. The Forests will "ratchet down" facilities when conditions warrant doing so, but fail to "ratchet up" when conditions improve.

- Roads should be a priority—Tribes need to be consulted directly, and they need fast and feasible access to deal with fire.

- The permitting process for outfitters needs to be user friendly— smoother procedure and better communication.

3. Stakeholders want to know their role and how their comments/ contributions are addressed. Parties are willing to participate in Forest Planning; they want to know how they can do so. They are concerned about transparency and accountability. How will their comments be considered? They look for follow-up and follow-through. Stakeholders and Tribes want to understand the decision space (what they can influence and negotiate) and know what roles they can expect to play in the process. They look for evidence that their participation matters; that their involvement is a good investment and contributes to the planning effort.

Sample statements illustrating this theme included:

- There is a perception that "the Forest Service does not have the data, does not know who its customers are, but makes decisions."

- There have been communication breakdowns—information goes nowhere—people will go to only so many meetings.

- Citizens perceive that policies are driven from top down—not derived from a local perspective.

- Be very clear on the purpose of meetings and information you are asking stakeholders to provide. Send background information out in advance so folks have time to review.

- Provide education and interpretation—perhaps develop a smartphone app for this.

- Provide a graphic of FPR (Forest Plan revision) basic steps—an accessible, understandable road map.

4. There is a general consensus that education is important. Many stakeholders stated that they did not know much (or anything) about Forest Plan revision. They wanted to be informed. Some parties voiced concern that the multiple-use mandate and mission of the Forest Service are not well understood. Listening session participants commented on the importance of education, from K-12 to involving the universities.

Sample statements illustrating this theme included:

- People need to understand the history and significance of grazing. Education and information are needed.

- Send out existing Forest Plan information or a link to the current Forest Plan—"You can't know where you are going until you know where you've been."

- Partner with the oil/gas community to offer educational tours.

- Help people understand the "what" and "why" of Forest Plan revision.

- The general public seems unaware—the Forest needs to do a better job of reaching the public.

5. Culture, history, and place attachments run deep in the areas of the Forests. A number of listening session participants commented on the importance of culture and history in the region of the Forests. They referred to sacred sites, special places, subsistence uses, struggling rural economies, and land rights. Some participants expressed concern that the people conducting the listening sessions—the Cadre members—were not local. Listening session comments highlighted the importance of Forest Service staff knowing the culture and history of the region, particularly in rural areas.

Sample statements illustrating this theme included:

- Socioeconomic impacts of current Forest policy and management decisions are strangling local rural economies and cultures. Timber mills are closing, local businesses are shutting down in rural communities because of campground/trail closures; there are fewer opportunities for firewood sales, and there are cuts in grazing allotments.

- Forest Service people need to be better trained; they need to better understand the local people. Rather than bringing in outside people, why not promote from within?

- There is a lack of knowledge of and respect for traditional cultures and uses, such as cutting firewood and grazing.

- Provide orientations for new agency personnel on local history and culture and increase sensitivity to local customs and culture throughout the agency.

6. Good communication is essential. Although not clearly or uniformly defined, "communication" was an important topic in the listening sessions. Cadre members asked about this topic directly and received a lot of comments. Participants said that the Forests' communicate with stakeholders well at times and on specific projects (e.g., the CFLRP) but need to make communication about forest management policies and practices (not just Forest Plan revision) a top priority. People wanted to be informed (as noted earlier) about National Forest activities, planning efforts, and regulations. People sought information about public meetings, open and closed areas, recreation activities, and special events. They hoped that the National Forest would communicate in both conventional ways (e.g., notices in the newspaper) and by using communication technologies (e.g., email and social media).

Sample statements illustrating this theme included:

- Communicate with local communities and business alliances when planning campground/road closures. Explain the reasons for closures before closures occur; provide public notice in local newspapers and via radio announcements.

- There is a low trust level with the public because of a history of ineffective communication, an inability or failure to cooperate/coordinate with partners to implement projects, inconsistency in how policies have been interpreted and applied between Forest Districts and between regional/Washington offices and the local Forests, and a failure to "close the loop" on commitments.

- Improve communication with stakeholders. Stakeholders want to know whom to contact for questions on topics such as recreation, timber, and road closures. Follow through with communications by calling stakeholders back.

- Communication could be better—a planning process begins with gusto and then fizzles out.

- Make information accessible and easy to find; the CFLRP is a good example.

7. Relationships and trust need attention. According to listening session participant comments, stakeholder relationships with the Forests ranged from very strong (e.g., with watershed councils) to very tenuous (e.g., some off-highway vehicle—OHV—groups). Some citizens and stakeholder organizations did not trust the Forest staff to manage the Forest for multiple use. For example, members of an off-road motorized recreation group expressed frustration that the National Forests did not listen to and respect their ideas during a travel management planning process. They were skeptical that contributions to Forest Plan revision would be considered.

Relationship quality, trust, and credibility were affected by management decisions, public participation processes, communication, and staff turnover. While citizens understood that people want to advance their careers, there were concerns that relationships between Forest personnel and communities suffer when key staff roles keep changing (e.g., the district ranger stays for two years and then leaves for another job at a different forest).

Sample statements illustrating this theme included:

- Explore ways to enhance the relationships with other state and federal agencies, such as the Natural Resource Conservation Service (NRCS). Create additional opportunities for face time.

- There is an inability to build and maintain good relationships between communities and the Forest because of staff turnover, employees not residing or "vested" within communities as a result of "zoned staffing patterns," reduction in face time with agency staff at all levels (i.e., the regional forester, the Forest supervisor, and the ranger district staff), and a lack of resources to get local on-the-ground projects completed in a timely manner.

- Indian tribes feel that the relationship with the Forest has been pretty good in the past—they would like to have more face-to-face consultation rather than via emails or phone calls.

- All District offices need to be staffed—doing so would improve communication and trust.

- The Forests have major credibility problems with the OHV (off-highway vehicle) community. OHV organizations do not want to sue (litigate) but have no other option.

8. Participation should be meaningful and consistent. The previous seven themes all relate to some degree to participation. The fact that more than 200 Tribal members and citizens participated in the listening sessions indicated that they considered participation in forest management an important investment of their time. Although a significant number of listening session participants did not know what to expect (or expected a more conventional public meeting), they shared ideas and frustrations regarding participation (or the lack thereof).

Sample statements illustrating this theme included:

- Create a safe environment for participation and the Forests' commitment to hold events in various parts of the Forests.

- Provide professional facilitation of public meetings.

- Be clear about where their ideas can contribute to the process and what they might influence (what is open for discussion and negotiation).

- Meetings and events should be held across the Forests with particular attention paid to rural areas. They should be accessible both in time and place.

- Getting out on the ground (e.g., field trips) fosters learning and understanding.

- Provide opportunities for written responses, especially for those who do not want to speak in front of large groups.

- Notify folks about upcoming meetings through email and flyers at grocery stores, post offices, and gas stations.

9. Don't just talk about participation and collaboration—demonstrate it. Citizens and Tribal members commented on collaboration. No single view emerged, and some listening session participants were skeptical. Just because the new Planning Rule calls for collaboration, one citizen noted, doesn't mean it will occur. Participants pointed out obstacles and constraints to collaboration while expressing the hope that the Forests would practice what the Planning Rule seemed to preach.

Sample statements illustrating this theme included:

- Maintain equal representation of all viewpoints. Everyone's voice carries the same weight. During travel management public meetings, professional associations/groups had people who were paid to attend while it was difficult for others to take off from work.

- Not all citizens have the advocacy skills that the professional organizations and groups bring to public meetings, and this will hinder their ability to articulate their points.

- Agencies are in conflict with one another—such as with minerals policy.

- A lot of the budget goes to fire.

- Some recreation groups (such as OHVers) want to volunteer to help the Forests but are ignored.

- The Forest Supervisor (headquarters) Office and local District staff need to be present at all meeting—they need to "hear and feel" what citizens have to say.

Conclusion

NEPA scholar Lynton Caldwell asserts that "the reconciliation of differences regarding the place of the environment in public policy thus became—and remains—a problem that is political, juridical, administrative, and at its base, ethical" (Caldwell, 1999, p. 2). Public participation as enshrined in NEPA, and subsequent public participation policies from local to global, attempt to reconcile and even integrate the four areas of environmental policy that Caldwell highlights. Public participation is political, attempting to model deliberative governance in contexts of imbalanced power (Simmons, 2007; Walker & Senecah, 2011). It is juridical, with stakeholders pursuing formal legal redress if they determine that more informal means of participation (such as collaboration) are insufficient (McKinney & Harmon, 2004; Nie, 2008). It is administrative, with directives such as the Forest Service's 2012 Planning Rule featuring the importance of engaging stakeholders through collaborative participation and communication. And participation is ethical, ideally consistent with principles of democracy, access, inclusion, and transparency (Fischer, 2003).

Although public participation varies among nations, agencies, and organizations, it is fundamentally about processes "by which public concerns, needs, and values are incorporated into governmental and corporate decision-making" (Creighton, 2005, p. 7). Public participation strategies and practices that stakeholders will likely respect and engage provide meaningful opportunities for influence, include access to relevant and understandable information, reach out to nontraditional and marginalized communities, and demonstrate to stakeholders where and how their ideas have been addressed.

As the introductory discussion of public participation throughout the world illustrates, nations recognize that public participation should be a component of environmental planning, decision making, and management. Yet public participation is often more powerful on paper than it is in practice, with some countries, particularly where democratic institutions are weak, falling short of providing citizens and stakeholders with a meaningful voice (Kakonge, 2006).

Why? There are many factors, such as capacity, corruption, and a lack of commitment. Consequently, public participation, while visible in planning policies throughout the world and at every scale, does not mean sustained action or innovation. The themes that have emerged from the comments of forest management listening session participants might be repeated and could apply in many countries, locales, and projects. Too often, natural resource managers may rely on a "command and control" approach to participation (Walker & Senecah, 2011) rather than asking stakeholders for their ideas and consequently listening closely to and acting on those ideas.

The listening sessions of the two Forests featured in this chapter provided stakeholder and Tribal members with opportunities to address matters of participation and communication in Forest Plan revision. Their comments highlighted concerns they had about their relationship with the Forests and their management personnel.

More importantly, their ideas—their experiences, concerns, interests, and questions—generated themes that can contribute to an organization's (such as an agency or ministry) development and implementation of a comprehensive and appropriate participation and communication strategy as part of project or planning work. The comments of listening session participants highlighted the importance of trust, acknowledgement, follow-through, learning, and, of course, listening.

NEPA and NEPA-like legislation, directives, procedures, and declarations may guide public participation in environmental policy decision situations, but they do not specify ways in which participation and communication should be enacted. Stakeholders can provide specifics—ideas about what is appropriate and feasible. By beginning at the grassroots level; by listening to and learning from involved and affected people, organizations and agencies—decision authorities—can scale up stakeholder engagement and improve public participation in environmental policy decision making, regardless of country or cause.

References

Andersson, K. (2013). Local forest governance and the role of external organizations: Some ties matter more than others. *World Development, 43,* 226–237.

Armitage, D., Berkes, F., & Doubleday, N. (2007). *Adaptive co-management: Collaboration, learning, and multi-level governance.* Vancouver, Canada: UBC Press.

Beierle, T. C., & Cayford, J. (2002). *Democracy in practice: Public participation in environmental decisions.* Washington, DC: RFF Press.

Burgess, H., & Burgess, G. (1996). Constructive confrontation: a transformative approach to intractable conflicts. *Mediation Quarterly, 13,* 305–322.

Butler, A., & Berglund, U. (2014). Landscape character assessment as an approach to understanding public interests within the European Landscape Convention. *Landscape Research, 3(39),* 219–236.

Caldwell, L. K. (1999). *The National Environmental Policy ACT: An agenda for the future.* Bloomington, IN: Indiana University Press.

Carey, M. P. (2013). *The Federal rulemaking process: An overview.* Washington, DC: Congressional Research Service.

Casperson, O. H. (2009). Public participation in strengthening cultural heritage: The role of landscape character assessment in Denmark. *Geografisk Tidsskrift— Danish Journal of Geography, 109*(1), 33–45.

Clarke, T. L., & Peterson, T. R. (2015). *Environmental conflict management.* Thousand Oaks, CA: Sage.

Cox, J. R. (2009). *Environmental communication and the public sphere* (2nd ed.). Thousand Oaks, CA: Sage.

Cox, J. R. (2012). *Environmental communication and the public sphere* (3rd ed.). Thousand Oaks, CA: Sage.

Council on Environmental Quality (2007). *Collaboration in NEPA: A handbook for NEPA practitioners.* Washington, DC: CEQ.

Creighton, J. L. (2005). *The public participation handbook: Making better decisions through citizen involvement.* San Francisco, CA: Jossey-Bass.

Daley, D. M. (2012). Public participation, citizen engagement, and environmental decision making. In M. E. Kraft & S. Kamieniecki (Eds.), *The Oxford Handbook of environmental policy* (pp. 487–503). New York: Oxford University Press.

Daniels, S. E., & Walker, G. B. (2001). *Working through environmental conflict: The Collaborative Learning approach.* Westport, CT: Praeger.

Daniels, S. E., & Walker, G. B. (2012). Lessons from the trenches: Twenty years of applying systems thinking to environmental conflict. *Systems Research and Behavioral Science, 29,* 104–115.

Dietz, T., & Stern, P. C. (Eds.). (2008). National Research Council. *Public participation in environmental assessment and decision making.* Washington, DC: The National Academies Press.

European Commission (updated 2016). The Aarhus Convention: The EU & the Aarhus Convention: In the EU Member States, in the community institutions and bodies. Retrieved from http://ec.europa.eu/environment/aarhus/legislation.htm

Fischer, F. (2003). *Citizens, experts, and the environment: The politics of local knowledge.* Durham, NC: Duke University Press.

Fisher, R., Ury, W., & Patton, W. (1991/2011). *Getting to yes: Negotiating agreement without giving in.* New York: Basic Books.

Graversgaard, M., Thorsøe, M. H., Kjeldsen, C., & Dalgaard, T. (2016, July). Evaluating public participation by the use of Danish water councils: Prospects for future public participation processes. Paper presented at the 12th European IFSA Symposium, Harper Adams University, Newport, England, UK.

Joint Summit Working Group (2009). *Achievements of the Summits of the Americas From Mar del Plata to Port of Spain.* Washington, DC: Organization of American States.

Kakonge, J. O. (2006). Environmental planning in Sub-Saharan Africa: Environmental impact assessment at the crossroads. Working Paper Number 9. New Haven, CT: Yale School of Forestry and Environmental Studies.

Kakonge, J. (2013). Improving Environmental Impact Assessment (EIA) effectiveness: Some reflections. Global Policy. Retrieved from http://www.globalpolicyjournal.com/blog/05/03/2013/improving-environmental-impact-assessment-eia-effectiveness-some-reflections

Kamphorst, D. A., Bouwma, I. M., & Selnes, T. A. (2017). Societal engagement in Natura 2000 sites: A comparative analysis of the policies in three areas in England, Denmark and Germany. *Land Use Policy, 61,* 379–388.

Kalen, S. (2009). The devolution of NEPA: How the APA transformed the nation's environmental policy. *William & Mary Environmental Law and Policy Review, 33,* 483–548.

Lindstrom, M. J., & Smith, Z. A. (2001). *The National Environmental Policy Act.* College Station, TX: Texas A&M University Press.

Long, R. J., & Beierle, T. C. (1999). *The Federal Advisory Committee Act and public participation in environmental policy.* Discussion paper 99–17. Washington, DC: Resources for the Future.

McKinney, M., & Harmon, W. (2004). *The western confluence: A guide to governing natural resources.* Washington, DC: Island Press.

Marara, M., Okello, N., Kuhanwa, Z., & Leentvaar, J. (2011). The importance of context in delivering effective EIA: Case studies from East Africa. *Environmental Impact Assessment Review, 31,* 286–296.

Nagendra, H., & Ostrom, E. (2012). Polycentric governance of multifunctional forested landscapes. *International Journal of the Commons, 6,* 104–133.

Natural England (2014). Landscape character assessment. Retrieved from http://www.naturalengland.org.uk/ourwork/landscape/englands/character/assessment/

Natural Resources Defense Council (2014). Environmental Issues: U.S. Law & Policy. Never eliminate public advice! Retrieved from http://www.nrdc.org/legislation/nepa-success-stories

Nelson, D. R., Folhes, M. T., & Finan, T. J. (2009). Mapping the road to development: A methodology for scaling up participation in policy processes. *Development in Practice, 19*, 386–395.

Nie, M. (2008). *The governance of western public lands: Mapping its present and future.* Lawrence, KS: University Press of Kansas.

Okello, N., Douven, W., Leentvaar, J., & Beevers, L. (2008). Breaking Kenyan barriers to public involvement in environmental impact assessment. *IAIA08 Conference Proceedings: The Art and Science of Impact Assessment.* 28th Annual Conference of the International Association for Impact Assessment, Perth, Australia.

Phillips, L., Carvalho, A., & Doyle, J. (Eds.). (2012). *Citizen voices: Performing public participation in science and environment communication.* Chicago: University of Chicago Press.

Schipper, E. L. F., Ayers, J., Reid, H., Huq, S., & Rahman, A. (Eds.). (2014). *Community-based adaptation to climate change: Scaling it up.* London: Routledge.

Schulte, W. J. (2013). Public participation still lacking from China's environmental laws. China Dialogue. Retrieved from https://www.chinadialogue.net/article/show/single/en/6482-Public-participation-still-lacking-from-China-s-environmental-laws

Senecah, S. L. (2004). The trinity of voice: The role of practical theory in planning and evaluating the effectiveness of environmental participatory processes. In S. P. Depoe, J. W. Delicath, & M.-F. Aelpi Elsenbeer (Eds.), *Communication and public participation in environmental decision making* (pp. 13–33). Albany, NY: State University of New York Press.

Simmons, W. M. (2007). *Participation and power: Civic discourse in environmental policy decisions.* Albany, NY: State University of New York Press.

Summit of the Americas Secretariat (2005). Follow-up and implementation: Mandates—democracy. Retrieved from http://www.summit-americas.org/sisca/dem.html

Swanwick, C. (2004). The assessment of countryside and landscape character in England: An overview. In K. Bishop & A. Phillips (Eds.), *Countryside planning: new approaches to management and conservation* (pp. 109–122). London, Earthscan.

Tudor, C. (2014). *An approach to landscape character assessment.* Core Document 40.20. Natural England. Retrieved from www.gov.uk/natural-england

USDA-Forest Service. (July 1997 & April 2000). *Collaborative stewardship within the Forest Service: Findings and recommendations from the National Collaborative Stewardship Team.* Retrieved from www.partnershipresourcecenter.org

USDI, Bureau of Land Management. (2003). Leaving a 4 C's legacy, a framework for shared community stewardship. Report to the Assistant Secretary of Land and Minerals Management on 4 C's principles, elements, barriers, projects & tools. Retrieved from www.blm.gov

Utembe, W. (2015). A critical appraisal of environmental impact assessment (EIA) in Malawi. *Malawi Journal of Applied Science and Innovation, 1*, 2–11.

Walker, G. B., & Senecah, S. L. (2011). Collaborative governance: Integrating institutions, communities, and people. In E. F. Dukes, K. E. Firehock, and J. E. Birkhoff (Eds.), *Community-based collaboration: Bridging socio-ecological research and practice* (pp. 111–145). Charlottesville, VA: University of Virginia Press.

Walker, G. B., Daniels, S. E., & Cheng, A. (2006). Facilitating dialogue and deliberation in environmental conflict: the use of groups in collaborative learning. In L. Frey (Ed.), *Facilitating group communication: Innovations and applications with natural groups* (pp. 205–238). Cresskill, NJ: Hampton Press.

Walker, G. B., Daniels, S. E., & Emborg, J. (2015). Public participation in environmental policy decision making: Insights from twenty years of collaborative learning fieldwork. In A. Hansen & R. Cox (Eds.), *The Routledge handbook of environment and communication* (pp. 111–130). London: Routledge.

Warbuton, D. (2004). Community involvement in countryside planning in practice. In K. Bishop & A. Phillips (Eds.), *Countryside planning: new approaches to management and conservation* (pp. 250–270). London: Earthscan.

Webler, T., Tuler, S., & Krueger, R. (2001). What is a good public participation process? Five perspectives from the public. *Environmental Management, 27*, 435–450.

World Bank Institute (2014). Multi-stakeholder collaborative action. Collaborative governance brochure insert. Retrieved from http://wbi.worldbank.org/wbi/Data/wbi/wbicms/files/drupal-acquia/wbi/multistakeholder_insert_fy12.pdf

Young, O.R. (2012). *On environmental governance: Sustainability, efficiency, and equity*. Boulder, CO: Paradigm Publishers.

Chapter 3

Rethinking Public Participation

The Case of Public Land Management in the American West[1]

Matthew McKinney

Introduction

Public land policy has been in the spotlight more than usual in recent years. On any given day, newspapers and social media in the American West highlight issues related to wildfire, recreational tourism, habitat connectivity, energy development, national monument designations, and calls for the transfer of federal lands to states. For the first time in a long time, public land policy was part of both Democratic and Republican presidential political party platforms in 2016.[2] And the Western Governors' Association 2016–2017 chairman's initiative, led by Montana Governor Steve Bullock, is focused on national forest and rangeland management (Montana Governor Bullock announces, 2016).

The current public and political attention to public land policy in the American West is neither surprising nor novel. Federal public lands (not including state or tribal lands) are one of the defining features of the American West and significantly influence the region's economies, communities, and culture (Stegner, 1969; Rasker, 2012; Gorte et al., 2012). They account for 28% of all land in the United States and 47% of the American West. More than 90% of all federal land is found in the 11 westernmost states and Alaska. The U.S. Forest Service and the Bureau of Land Management

(BLM) administer about 34% of the western landscape, including almost 85% of Nevada; more than 50% of Idaho, Utah, and Oregon; and more than 40% of the land in four other western states.

Given the prominence of public lands in the West, it is not surprising that Westerners have debated the appropriate use, management, and even ownership of these lands since they were first established (Kemmis, 2001). This debate played out in different ways from 1900 through the 1960s, reemerging in the 1970s under the banner of the "Sagebrush Rebellion," in the 1990s as the "county supremacy movement," and most recently as the federal lands "transfer movement." (Leshy, 1980; Alexander & Gorte, 2007; Keiter & Ruple, 2015).[3]

This historic narrative not only highlights enduring tensions and acute conflicts over public land management, but also reveals that debates over public land policy revolve around three related sets of questions (Wondolleck, 1988; Silken, 2009). First are questions of purpose—for example: What are the objectives, priorities, or uses for which public lands should be managed? How should resources be allocated? Second are questions of process—for example: Who makes what decisions? And what role do citizens, stakeholders, experts, and local elected officials play in making decisions and implementing outcomes? The third set of questions revolves around jurisdiction, particularly the question of whether the federal government should retain ownership and management or whether there are better alternatives. These questions overlap each other because those who control the decision-making process determine what constitutes acceptable uses. In this respect, public lands policy and the practice of democracy are fundamentally linked. As explained by Kemmis (2001), public lands exemplify democracy in two important ways: by allowing equal access to federal lands and resources for all Americans and by including all Americans in the decision-making processes that determine how these lands are managed.

In light of these geographic, historic, and political imperatives, one argument of this chapter is that we are not likely to effectively resolve issues of purpose and policy until we create more effective democratic processes to bring together people representing diverse interests with the best available information. The conventional approaches to public participation on public land management—as mandated by the Administrative Procedures Act (1946), National Environmental Policy Act, or NEPA (1969), Federal Land Policy and Management Act (1976), and National Forest Management Act (1976)—revolve around two basic objectives: to "inform and educate" citizens and to "seek their input and advice." As explained later, even more recent laws, administrative rules, and policies that encourage or mandate some type of collaboration fall under these two basic objectives. While these objectives, and the methods that support them, are valuable, they compel public land management agencies to serve as a kind of ringmaster in a field of competing interests. Given the design of the decision-making system, where the agency is solely responsible for the weighing and balancing of trade-offs and making decisions, the different "publics" are increasingly unencumbered from any responsibility to help solve problems.

In *A Conspiracy of Optimism*, Paul Hirt (1996) suggests that this approach to public participation and decision making seems to encourage an adversarial approach to public participation by more or less promising that all parties can get what they want, instead of creating the conditions necessary to bring everyone to the table to share the responsibility of solving problems by working together. The process is perhaps best represented in Arnstein's classic "ladder of citizen participation" (1969) as "degrees of tokenism," with perhaps a shade of "partnership." The outcomes are well known to people who live, work, and play on public lands. While agencies do their best to balance competing interests and make decisions on the best available science, the entire process often leaves citizens, advocates, and decision makers dissatisfied with the outcome. This dissatisfaction in turn leads to a recurrence of disputes, which strains relationships and increases transaction costs.

In contrast to these conventional approaches to public participation and decision making, a variety of innovative approaches to public engagement and shared problem solving are emerging within communities, watersheds, and larger landscapes. Often referred to as the "collaboration movement," these innovations started to appear in the early 1990s when citizens and stakeholders became frustrated and dissatisfied with more conventional, government-driven processes to manage public lands (Brick et al., 2001). These homegrown, grassroots processes tend to be citizen driven and place based. In most cases, they do not have any official authority. Instead, they generate legitimacy, credibility, and effectiveness by building broad-based coalitions or a "constituency for change." So-called "coalitions of the unalike" (Snow, 2001) create public processes that are inclusive, informed, and foster a sense of shared ownership for the process, decisions, and outcomes. And they are achieving notable outcomes in terms of economic development, community vitality, and environmental stewardship.

Even at the Malheur National Wildlife Refuge in Oregon, where a high-profile occupation of federal lands occurred in 2016 in protest of federal control of western lands, there has been a quiet yet effective, collaborative group working for more than ten years. Known as the High Desert Partnership (see www.highdesertpartnership.org), this coalition of ranchers, environmentalists, and government agencies has facilitated a process of listening, learning, and cooperative land-use planning. According to recent reports, it has transformed local land-use politics from a state of gridlock dominated by acrimony and litigation to one of implementing innovative solutions to complex problems by working together, reaching agreements, and rebuilding the sense of community (Brown, 2016).

Participants in this partnership apparently did not welcome the protestors, believing that their homegrown approach to working with the federal land management agencies is more constructive and promising. One of the partnership's first accomplishments was to create a 15-year Comprehensive Conservation Plan for the refuge where—among other provisions—grazing permits are issued every five years rather than annually, and cattle are kept in higher pasture later into the summer to allow the chicks of Sandhill Cranes, Bobolinks, and other birds to hatch in the wet meadows (Marris, 2016). As explained more fully below, this is not an isolated case of collaborative conservation, but rather another example of what is fast becoming the norm in public land management.

In spite of the negative press and political rhetoric about the problems facing public land management, there actually is a great deal of cooperation and innovation occurring to solve problems, build trust, and sustain both communities and landscapes in the American West. The challenge, or better yet, the opportunity, is to rethink our conventional approaches to public participation and governance by integrating the lessons of these more informal, collaborative processes into formal decision-making processes. This is not a call for agencies to abdicate their decision-making authority, but for them to share responsibility for solving problems.

In this respect, it is important to distinguish governance from government. Government occurs when people with formal legal authority make plans and take action. By contrast, governance is what happens when citizens and groups (often including government agency officials) work together to plan and act based on their shared goals. Such efforts may or may not have formal authority or power (Bingham, Nabatchi, & O'Leary, 2015). Governance refers to the style or method by which decisions are made and the way in which conflicts among actors are resolved. Governance is about representation, style of interaction, authority, and decision rules. It also refers to processes that support governance: that is, fostering scientific and public learning as well as building civic and political will.

Homegrown Innovation:
The Case of the Crown of the Continent

To illustrate the evolution of innovative approaches to public participation and shared governance, consider the case of the Crown of the Continent (COTC). The COTC is an 18-million–acre transboundary ecosystem that includes parts of Montana, Alberta, and British Columbia.[4] It is an ecological crossroads where plant and animal communities from the Pacific Northwest, eastern prairies, southern Rockies, and boreal forests mingle. The majestic spine of mountains is the headwaters for North America, where pristine rivers originate and flow to the Pacific Ocean, Gulf of Mexico, and Arctic Ocean. The COTC is one of very few landscapes on the continent that retains its full complement of native habitat and native predators—wolves, grizzly and black bears, cougar, coyote, fox, wolverine, bobcat, and lynx—as well as large populations of moose, elk, bighorn sheep, pronghorn, and deer.

The COTC is, and has been, home to a number of indigenous people. Ancestors of the Blackfeet, Kainaiwa, Ktunaxa, Salish, and Kootenai peoples were among the first to hunt, fish, and gather plants for food and fiber here. By the early 1800s, when the first white explorers and trappers arrived, much of the region was already settled, with tribal territories, hunting grounds, and travel routes well established. As the population grew, some people saw development as a threat to the region's natural heritage and beauty. In the late 1890s, several people, including the editor of *Forest and Stream* magazine—George Bird Grinnell—lobbied Congress to establish a national park south of the Canadian border. In a series of articles, Grinnell referred to the region as the "Crown of the Continent." A forest preserve was set aside in 1897, but the area remained open to mining and logging. Grinnell and other conservationists continued promoting the area's unique features, and finally, in 1910, President Taft signed a bill creating Glacier National Park, which borders Waterton Lakes National Park, created by Canada in 1895.

Local Rotary Clubs in Alberta and Montana rallied around the idea of this shared landscape, and in 1932, the governments of both Canada and the United States voted to designate the parks as Waterton-Glacier International Peace Park—the world's first such designation. UNESCO named Glacier

National Park as a Biosphere Reserve in 1976 and recognized Waterton Lakes with the same designation in 1979. Comprising about 1.3 million acres, the two parks were named a World Heritage Site in 1995, acknowledging the area's rich ecological and cultural values. In Montana, about 1.6 million acres of federally protected wilderness extend around Glacier National Park. Several areas throughout the transboundary ecosystem benefit from additional special conservation designations, including wild and scenic rivers, provincial parks, wildlife management areas, and recreation areas. Many additional acres of private working landscapes are protected under conservation easements.

Thanks to this remarkable history of stewardship, the COTC endures today as a natural oasis in an increasingly developed world. Like many large landscapes in the American West, however, the COTC is currently faced with a number of issues related to climate change, water resources, wildlife corridors and habitat conservation, evolving economic opportunities, and patterns of growth and development (McKinney & Johnson, 2013). In response to these complex issues, individuals and organizations through-out the COTC are creating new forms of public engagement and shared problem solving—what might be referred to collectively as an "ecology of governance." In a formal sense, the COTC includes two nations, two provinces, one state, and seven tribes and First Nations, with more than 20 government agencies exercising some type of authority and management on the landscape. Although the landscape is jurisdictionally fragmented, each of these institutions plays an important role in managing natural resources. Unfortunately, the most compelling issues facing the COTC, from invasive species to weeds to wildlife corridors, wildfire, water, and so on, present themselves at a spatial scale that crosses jurisdictional and cultural boundaries.

While legal and institutional boundaries delineate ownership and management authority, they also create barriers among neighbors and can reinforce disparate cultures, attitudes, goals, and values. In spite of these challenges, people who care about the COTC and its future are creating new opportunities for public engagement and shared problem solving. What is occurring, in fact, is a nested system of collaborative arrangements that are similar, at least in part, to Ellinor Ostrom's notion of "polycentric systems of governance" (2009). Today, more than a hundred agencies and community-based partnerships are working to promote and support livable communities, vibrant economies, and healthy landscapes within the COTC.

Starting at the smallest geographic scale, there are at least 20 commu-nity-based partnerships in the COTC, most of them initiated and convened by citizens. These community-based partnerships create the basic building blocks within the nested system of governance (Weber, 2003). Consider, for example, the Blackfoot Challenge (Blackfoot Challenge, 2015). This land-owner-led nonprofit organization coordinates management of the Blackfoot River, its tributaries, and adjacent public and private lands, a total of about 2,400 square miles. It is organized locally and known nationally as a model for preserving the rural character, ecological health, and natural beauty of the watershed. The mission of the Blackfoot Challenge is to coordinate efforts that enhance, conserve, and protect the natural resources and rural lifestyles of the Blackfoot River Valley for present and future generations. It supports environmental stewardship through cooperation of private and public interests. Private landowners, federal and state land managers, local government officials, and corporate landowners compose the membership of the Board of Directors. Although the Blackfoot Challenge does not have any formal, legal authority to manage lands and resources in the watershed—either public or private lands and resources—the participants share a common vision and belief that successful land, water, and natural resources management is most likely to result from building trust and sharing responsibility.

As presented in Figure 3.1 on page 76, the Blackfoot Challenge has achieved several notable accomplishments since it was created in 1993. It also illustrates how community-based partnerships in the COTC "nest" alongside each other to help manage the land, water, and people within distinct yet adjacent watersheds. To complement these basic building blocks, at least nine independent initiatives have emerged since 1994 to promote and support shared problem-solving at the scale of the transboundary COTC. While none of these initiatives has any formal authority to make and implement decisions, each plays a critical role in the ecology of governance—exchanging information, building relationships, and creating opportunities to work together. Along with the community-based partnerships such as the Blackfoot Challenge, they help build the civic and political will to address complex natural resource and related issues that cannot be effectively addressed by any single community, stakeholder group, or government agency.

The Crown Managers Partnership (CMP), for example (www.crown-managers.org), emerged in 2001 as an interagency forum for about 20 land management agencies in Montana, British Columbia, and Alberta. This voluntary partnership provides a forum for management agencies to identify common needs and interests, develop joint initiatives, and leverage resources as appropriate. It convenes an annual public forum to examine both ongoing and emerging issues and to inform decision makers at all levels on priority issues and actions. It is important to emphasize that, like

Keeping Landscapes Working
- ❖ 150,000 acres under conservation easement, available for agriculture and wildlife
- ❖ 238,000 acres of corporate timber land kept working in conservation status
- ❖ 83% of watershed in conserved status (private land conservation easement or public ownership)
- ❖ 41,000 acres of public and private land managed by community council

Reducing Conflicts
- ❖ Keeping grizzly bear conflicts below 94% since 2003 and reducing wolf conflicts since 2008

Conserving Water for Agriculture and Fish
- ❖ Conserving 10s of millions of gallons of water in a typical drought year through voluntary plans
- ❖ 50% of the irrigation systems participating in the irrigation/energy efficiency program

Making Communities Safe and Maintaining Forest Health
- ❖ Reducing forest fuel loads on an average of 500 acres each year since 2009

Connecting Classrooms & Communities with Place-Based Education
- ❖ Educating 500 youth each year since 1993
- ❖ Reaching 1,500 adults each year since 2004

Transferring the lessons learned through community-based conservation
- ❖ Hosting the first America's Great Outdoor events in the Nation on June 1, 2010
- ❖ Approval for a private landowner advisory group to Secretaries of Interior and Agriculture
- ❖ Formation of Partners for Conservation to support community-based conservation across America
- ❖ Model for new National Fish and Wildlife Foundation Landscape Conservation Stewardship program

Managing Noxious Weeds Across Fence Lines
- ❖ Managing an average of 1,000 acres each year since 2000

Figure 3.1. Blackfoot Challenge Accomplishments.

the Blackfoot Challenge and other community-based partnerships, the work of CMP is nonbinding; it depends on the participating agencies going back to their particular jurisdictions and implementing projects consistent with agreed-on objectives and strategies.

According to CMP's *Strategic Conservation Framework 2016–2020* (Crown Managers Partnership, 2016), its major accomplishments over the years include creating and maintaining a transboundary data base on the status and trends of ecological health within the ecosystem; preventing the spread of aquatic invasive species; increasing the resilience of native, cold-water salmonids; and managing noxious weeds. These and other accomplishments demonstrate CMP's commitment to overcoming a variety of barriers (legal, financial, organizational, information, and so on) to the management of natural resources problems that cut across national and agency jurisdictional boundaries.

Realizing that the future of the COTC is being shaped by more than 100 government agencies, nongovernment organizations, and community-based partnerships, the Center for Natural Resources & Environmental Policy (based at the University of Montana) and the Center for Large Landscape Conservation, in partnership with several other organizations, launched the Roundtable on the Crown of the Continent (www.crownroundtable) in 2007. Prior to the Roundtable, the various initiatives operated largely independent of each other; people were connected to the landscape but not to each other. The Roundtable seeks to fill this gap by providing an ongoing forum to bring together individuals and organizations that care about the region.

Through workshops, forums, policy dialogues, conferences, and online newsletters, the Roundtable provides an independent, nonpartisan forum to exchange ideas, build relationships, and explore opportunities to work together. A leadership team that includes representation from community-based partnerships, nongovernmental organizations, communities, tribes and First Nations, agencies, and other people who care about the COTC governs the Roundtable. In 2016, the Roundtable received the *Climate Adaptation Leadership Award for Natural Resources* by the U.S. Department of the Interior for "catalyzing a landscape-scale, collaborative approach to the conservation of natural resources and adaptation actions across 18 million acres in Montana, Alberta, and British Columbia" (U.S. Department of Interior, 2016a). The award recognizes the Roundtable's adaptive management initiative, which harnessed financial and other resources and then invested the resources at various spatial scales and across multiple sectors to facilitate climate change adaptation.

Moving on and scaling up even further from the level of the COTC is the Yellowstone to Yukon Conservation Initiative (Y2Y), an effort to protect wildlife habitat and corridors across a 500,000-square-mile landscape—nearly three times the size of California (www.y2y.net). Y2Y began as a network of biologists and conservationists concerned about wildlife and their habitat. Today the organization focuses on protecting key connectivity areas for wildlife that are threatened by habitat loss, invasive species, and climate change. Y2Y also works closely with private landowners, community leaders, and others to address a range of issues related to land use, community and economic prosperity, and wildlife management. In 2015 alone, Y2Y protected more than 250,000 acres of land in Alberta's Castle watershed, ensured protection for 14 million acres of land in Canada's Yukon Territory, provided technical and facilitative support to the efforts of 118 partners to enhance collective impact in the Yellowstone to Yukon region, collaborated on 67 conservation projects that protect habitat and connect wildlife throughout the region, and raised $326,000 to support grassroots projects (Y2Y, 2016).

One of the most recent additions to the ecology of governance in the COTC is the Great Northern Landscape Conservation Cooperative (GNLCC) (www.greatnorthernlcc.org). This initiative, led by the U.S. Fish and Wildlife Service and other federal agencies, is developing scientific capacity to address climate change and other stressors to wildlife species and habitats within the Northern Rockies and the Columbia River Basin. The Cooperative provides scientific and technical support to government agencies, including tribes and First Nations, in part to support adaptive management and large-landscape conservation. The accomplishments of this cooperative initiative are impressive and include a variety of studies, tools, and projects to advance large-landscape conservation, in addition to building the capacity of several partners and creating a five-year transboundary science plan (GNLCC, 2015).

Several other homegrown initiatives further illustrate the variety of innovative approaches to public participation and shared problem solving emerging in the COTC. In response to a growing national debate over the use of mountain bikes in wilderness areas, Montana High Divide Trails formed in 2007 to find common ground to inform agency decision makers (Klugin, 2016; Wild Montana, n.d.). The group includes representatives from the Montana Wilderness Association, a local land trust, several mountain biking clubs, a backcountry horseman's association, and others. This "coalition of the unalike" has found that the U.S. Forest Service is likely to listen to it when it offers consensus recommendations on travel plans and the like. Its accomplishments include a network of "front country" foothill trails located near Butte and Helena that provide bicyclists, horse enthusiasts, hikers, and other quiet users access to mountain trails and open space linked to the rugged Continental Divide trail system.

Another example is the Innii Initiative (Locke, 2015; Innii Initiative, n.d.). In September 2014, members of the Blackfeet Nation, Blood Tribe, Siksika Nation, Piikani Nation, Fort Belknap Reservation, Fort Peck Reservation, Confederated Salish and Kootenai Reservation, and the Tsuu T'ina Nation signed the "Northern Tribes Buffalo Treaty." The intent of this transboundary treaty—the first such treaty among these disparate tribes in more than 150 years—is to bring wild buffalo back to tribal lands to restore its cultural, spiritual, nutritional, and ecological roles. In April 2016, nearly 100 bison were reintroduced to the Blackfeet Indian Reservation in northern Montana from Elk Island National Park in Alberta (Chaney, 2016).

Replicating Innovative Approaches

The ecology of governance in the COTC illustrates a growing trend in public participation and shared problem solving—a trend where citizens, nongovernmental organizations, universities, and other associations are taking the initiative to catalyze, convene, and coordinate public forums to exchange information, solve problems, and implement solutions. In most cases, these homegrown forums are designed to supplement, not replace, formal decision-making systems. In some cases, they allow the formal decision-making processes to work better—such as when the groups involved in Montana High Divide Trails find common ground and offer consensus recommendations to the U.S. Forest Service. This type of supplemental civic engagement does not replace the public participation required by the U.S. Forest Service prior to making decisions and taking actions, but it often informs that decision process, reduces the amount and intensity of conflict, and helps generate durable solutions that can be implemented on the ground.

This trend not only suggests a shift from an expert-driven model of decision making to more democratic approaches, but also raises some important questions about "governance" and the role of citizens, professionals, and communities in decision making. From a political perspective, this trend in homegrown, civic engagement creates a healthy tension between bottom-up and top-down approaches to governance. In *Planning with Complexity*, Judith Innes and David Booher (2010) suggest that this tension can be explained, at least in part, by the difference between "instrumental rationality" and "collaborative rationality."

Instrumental rationalists tend to approach natural resource issues as largely technical problems that can be effectively resolved by the best available science and the separation of politics from decision-making. This model emerged during the progressive era around the 1900s and continues to serve in large part as the foundation for public land management agencies (Hays, 1959; Nelson, 1995). By contrast, collaborative rationality sees the world as inherently uncertain and assumes that all decisions are necessarily contingent. From this perspective, planning and policy are not about finding the best solution (indeed, there is not likely to be one best solution), but rather discovering ways of proceeding that are better than the status quo. Most federal land management agencies, as well as others, seem to embrace this inherent uncertainty in natural resource decision making—at least to some degree—as demonstrated by the growing practice of adaptive management and planning (Schultz & Nie, 2012; Allen & Garmestani, 2015).

Public processes characterized by collaborative rationality engage diverse members of a community, including citizens, stakeholders with diverse needs and interests, as well as experts and agencies. They work together to jointly learn and generate solutions in the face of conflict, changing conditions, and conflicting sources of information. Such processes, as illustrated by the High Desert Partnership, the ecology of governance in the COTC, and the plethora of large-landscape conservation initiatives in the Rocky Mountains (McKinney & Johnson, 2013), not only generate new ways to move forward but also help communities adapt and become more resilient in the face of new challenges. In other words, the successful practice of collaborative governance within communities, watersheds, and ecosystems builds social, political, and intellectual capital that can then be applied to other issues facing individuals, groups, and communities. In this respect, collaborative governance is changing the political culture of the places and communities where it is practiced.

In light of these innovative trends in public participation and shared problem solving, how can we rethink the more conventional approaches to public participation and shared problem solving by integrating the lessons of the more informal, collaborative processes into formal decision-making processes? (Kemmis & McKinney, 2011). There seem to be two general responses to this question: first, to foster innovations within the existing legal and institutional system; and second, to begin experimenting with alternatives to the established decision-making system. Let's start with this latter option.

Beginning in the mid- to late 1990s, several observers started calling for a series of pilot projects or experiments in governance as a way to foster more innovative (and effective) approaches to public participation, decision making, and stewardship on public lands (Kemmis, 2001). Taken as a whole, the idea was to foster a diverse portfolio of experiments on public land governance—similar, in part, to the idea of a diversified portfolio in the investment world.

- In 1999, a broad-based group of participants came together to test the hypothesis that collaborative processes could and should be more effectively integrated into the decision-making process governed by NEPA (Center for the Rocky Mountain West, 2000). Among other things, they called for pilot projects to test the possibilities and limits of collaboration, including the degree to which decision-making authority might be vested in collaborative groups.

- A different group meeting in the late 1990s, referred to as the Forest Options Group, suggested a collaborative governance option, where a national forest plan would be written and the forest supervisor hired by a local board of directors (Forest Options Group, 1998). The participants would be required to follow all environmental laws but would be allowed to depart from internal agency procedures for the purposes of making management decisions.

- Still another broad-based group, meeting at Lubrecht Forest outside Missoula, Montana, in 1998, recommended the creation of a new Region 7 of the U.S. Forest Service. The original Region 7 was absorbed into two other regions in 1966 and the regions were never renumbered, so there has not been a Region 7 for decades (Snow et al., 1998). The new Region 7 would be a "virtual region" consisting of a diverse portfolio of pilot or experimental forests. Like the other proposals, it would include an opportunity for management plans to be written and implemented by a local collaborative group.

- More recently, Professor Robert Nelson (2015) has called for a series of charter forests. Much like charter schools, the key principle of charter forests is freedom with accountability. Charter forests would be freed from the centralized administration of

the Forest Service, and management would devolve to auton-
omous forests capable of more creative and locally responsive
management.

The common theme in all four of these proposals is that they would
turn planning and management—not ownership—over to community-based
partnerships, something like the Blackfoot Challenge. Just as "inside the box"
innovations (which are addressed below) allow the agencies to demonstrate
their willingness and capacity to incorporate collaborative methods within
established procedures, these community-based collaborative experiments
would give diverse groups of stakeholders a chance to prove they are capable
of ecologically sustainable stewardship of public lands.

Within the past two years, there have been additional calls for similar experiments in comanagement or what Kirk Emerson, the founding
director of the U.S. Institute for Environmental Conflict Resolution, refers
to as "collaborative federalism," where joint decision making occurs among
multiple governing units in contrast to divided and distributed decision
making (Emerson & Murchie, 2010; Emerson and Nabatchi, 2015). Native
Americans, conservationists, and other stakeholders in the proposed Bears
Ears National Monument in southern Utah (Bears Ears Tribal Coalition,
2015) and the Badger-Two Medicine sacred area in the COTC (Weaver,
2015; Dax, 2016) have called for opportunities to jointly manage these
landscapes. On December 28, 2016, President Obama created Bears Ears
National Monument in southeast Utah (White House, 2016). To help guide
the management of the monument, the proclamation creates the Bears
Ears Commission, composed of representatives from the Hopi Nation,
Navajo Nation, Ute Mountain Ute Tribe, Ute Indian Tribe of the Uintah
Ouray, and Zuni Tribe.

The limitation of these suggestions for experiments in public land governance is that they require either the president or Congress—or both—to
create the legal and institutional space to experiment with different models
of governance. While many people support this strategy, Congress currently
seems to be more focused on rolling back President Obama's environmental
achievements, promoting energy development on public lands, and transferring decision-making power, if not outright ownership, to the states through
various mechanisms (Conference of Western Attorneys General, 2016; Keiter
& Ruple, 2015). It is important to emphasize that these calls for a portfolio of experiments in public land governance are completely different from
current calls for the transfer of federal lands to the states.

While the call for a series of pilot projects or experiments in governance may grow out of a frustration with the existing federal land management system, the arguments in support of pilot projects recognize the fundamental effectiveness of homegrown, innovative, collaborative approaches to public lands management and seek to create legal and institutional space to replicate these types of arrangements as a matter of public policy. They recognize the value of sharing responsibility to solve public land problems, not by shifting who owns federal lands, but by working together across political, jurisdictional, and other boundaries. Rather than building on this growing legacy of sharing responsibility and problem solving in the Rocky Mountain West and elsewhere, the current political debate revolves around a winner-take-all approach to policy and governance.

In addition to continuing to advocate for a diverse portfolio of experiments in shared problem solving, the second approach to replicating the spirit and dynamics of collaborative governance is to foster innovation within the existing legal and institutional framework. Three examples illustrate how public land management agencies are moving in this direction.[5]

The first example is perhaps best described as an example of how the existing legal and institutional system can be improved through a pilot project authorized by Congress. Congress passed the Collaborative Forest Landscape Restoration Program (CFLRP) in 2009 (Omnibus Public Land Management Act, 2009). The intent of the statute is to authorize a limited number of projects to accelerate restoration on high-priority landscapes, support economic stability in rural communities, and reduce the risk and associated costs of catastrophic wildfire. An advisory committee overseeing implementation of this program selected projects on the basis of these goals and criteria. Projects were also selected on the strength of their collaborative capacity, demonstrated first and foremost by the mix of individuals and organizations that prepared the proposals. In other words, the CFLRP created the right set of incentives for people with diverse needs and interests to come together and forge a common vision and strategy (Bixler & Kittler, 2015). According to the program's five-year report (U.S. Forest Service, 2015a), the 10 pilot projects have generated the following accomplishments:

- More than 1.4 million acres treated to reduce the risk of catastrophic fire

- More than 84,570 acres of forest lands treated to achieve healthier condition through timber sales

- More than 1.33 million acres improved for wildlife habitat

- More than 73,600 acres treated for noxious weeds and invasive plants

- More than 1,256 million board feet of timber volume sold

- More than $661 million in local labor income

- An average of 4,360 jobs per year

CFLRP projects have also attracted new partners and built community relationships, leveraging more than $76 million in matching funds. By most metrics, the CFLRP seems to be a good example of how to integrate the "secret sauce" of collaborative governance into the existing legal and institutional framework governing public land management.[6]

The second example is the new planning rule adopted by the U.S. Forest Service (U.S. Forest Service, 2012). In 2012, after working through a multiparty collaborative process, the agency adopted new administrative rules to guide the process of revising and updating land management plans.[7] Among other things, the 2012 planning rule directs the U.S. Forest Service to "engage the public . . . early and throughout the planning process . . . using collaborative processes where feasible and appropriate . . . [as well as] the full spectrum of tools for public engagement . . ." Compared to the 1982 planning rule, the 2012 planning rule emphasizes collaboration, requires improved transparency, and strengthens the role of public participation and dialogue throughout all stages of land management planning while also retaining the traditional notice and comment procedures under NEPA (U.S. Forest Service, 2012, 2015b, 2016).

In the fall of 2014, the Center for Natural Resources & Environmental Policy at the University of Montana was asked to document and evaluate lessons learned with respect to public participation and collaboration in the 12 "early adopters" of the 2012 planning rule—the first national forests to revise and update their land management plans under the new rule (Center for Natural Resources and Environmental Policy, 2015). Based on that evaluation, several national forests are employing what might be considered "best practices" in collaborative planning, including but not limited to:

- Using professional facilitators to help design and guide the public process

- Dedicating a U.S. Forest Service staff person to serve as a "collaboration specialist" to help guide the public participation process

- Completing stakeholder assessments up front to clarify the needs and interests of individuals, groups, and communities; and to explore how they want to be involved in the process

- Jointly preparing public participation plans based on the stakeholder assessments

- Engaging the public prior to initiating the environmental analysis required by NEPA

- Using participatory mapping tools, which allow people with diverse interests to jointly identify areas suitable for wilderness designation, timber harvesting, and other resource uses. In addition to providing spatially relevant information, this type of interactive exercise allows individuals and groups to exchange ideas with each other and Forest Service officials, to consider potential conflicts and trade-offs, and to otherwise build and enhance relationships.

A limited number of national forests have gone even further. In the Nantahala and Pisgah National Forests in North Carolina, three different stakeholder groups attempted to create a multiparty collaborative process to run alongside, feed, and otherwise supplement the planning process. Each of these processes failed to generate sufficient momentum in large part because the self-appointed stakeholder groups limited who could participate. As a result, the National Forest Foundation was asked to step in and convene a single, more inclusive collaborative process to provide input and advice to the Forest Service as the planning process unfolds (see www.nationalforests.org/stakeholdersforum). This single collaborative group is up and running, and according to one person close to this process, the Nantahala & Pisgah National Forests are working hard to manage an open, transparent, and collaborative process that fosters shared problem solving (McKinney, 2016b). Apparently the National Forests practice what they refer to as "radical transparency" with the stakeholder forum as well as the larger public engagement process. They have opened up their interdisciplinary team meetings to observers and release draft sections of the plan online as they are completed.

In the Flathead National Forest in Montana, a diverse collection of individuals and groups created the Whitefish Range Partnership to seek agreement on recommendations for this particular area (Whitefish Range Partnership, 2013; McKinney, 2016a). Representatives of wilderness, timber, motorized and nonmotorized recreation, and the local communities worked alongside Forest Service officials and arrived at a set of consensus recommendations on land use and management for the Whitefish Range. These recommendations were rolled into the proposed action to initiate the NEPA process. This innovative approach to public participation and shared problem solving did not violate the Federal Advisory Committee Act because the Forest Service did not convene the partnership, and other people had a similar opportunity to provide input and advice prior to the start of the NEPA process (Public Policy Research Institute, 2006).

In other work with different national forests, the Center for Natural Resources and Environmental Policy has suggested a similar innovative approach. Rather than creating new collaborative partnerships for national forest planning, the Center suggested that national forests should build on existing community-based partnerships. In the case of the Helena and Lewis & Clark National Forest, for example, there are about ten different multiparty collaborative partnerships, all functioning with a track record of success. Given that these partnerships have done the heavy lifting of bringing diverse interests and viewpoints to the table, building trust, and achieving results on the ground, they could provide a solid foundation for public participation during the planning process. While they would not be a substitute for other opportunities for public participation, such well-established partnerships could help convene and facilitate public forums on issues related to an emerging forest plan. In many cases, these types of community-based partnerships are already working with the Forest Service to implement projects, so in part this is an opportunity to move from collaborative implementation to collaborative decision making.

These and other examples demonstrate how the 2012 planning rule has provided the legal and institutional space for the U.S. Forest Service to experiment with some innovative approaches to public engagement and shared problem solving. Not all of the national forests currently updating their land management plans via the 2012 planning rule have taken advantage of this opportunity. Nevertheless, realizing that these experiments are not perfect and that we do not yet know their final impact, the U.S. Forest Service should be commended for going above and beyond the conventional approach to public engagement as defined in the National Forest Management Act and NEPA.

The third and final example of fostering innovative public participation and shared problem solving within the existing legal and institutional framework for public land management was the proposed planning rule for the BLM (U.S. Department of Interior, 2016b). According to the summary of the proposed rule (known within the agency as Planning 2.0), the new framework would allow the BLM ". . . to more readily address landscape-scale resource issues, such as wildfire, habitat connectivity, or the demand for renewable and non-renewable energy sources and to respond more effectively to environmental and social changes . . . emphasize the role of science in the planning process and the importance of evaluating the resource, environmental, ecological, social, and economic conditions at the onset of planning . . . affirm the important role of other Federal agencies, State and local governments, Indian tribes, and the public during the planning process, and would enhance opportunities for public involvement and transparency during the preparation of resource management plans . . . clarify existing text and use plain language to improve the readability of the planning regulations" (U.S. Department of Interior, 2016b).

Planning 2.0 was developed over three years with extensive input and advice from a broad cross-section of individuals and organizations interested in and affected by management of BLM lands and resources. Contrary to planning rules adopted in 1979, 1983, and 2005, Planning 2.0 sought to improve public participation in resource management planning by involving the public earlier in the planning process, to increase public participation across political jurisdictions, and to arrive at more collaborative decisions around the use and conservation of lands and resources managed by the BLM.

While these proposals might seem like common sense, not everyone agrees. The Western Governors' Association argues, among other things, that the BLM proposes to shorten public comment periods in two important steps of the resource management planning process—during the development of the plan and the review process (Mead, 2016). WGA explains that "[a]ny process that reduces BLM's responsibility to actively inform the public of its actions represents a retreat from openness and transparency." The National Association of Counties argues that the new opportunity for public participation during the "planning assessment" phase of the process appears to give unelected special interests an equal seat at the table with local and state officials, which effectively places the views of sovereigns among the crowd of public and stakeholder views (U.S. House, 2016).

By contrast, the Missoula County Commission believes the proposed rule provides "additional opportunities for public involvement earlier in the planning process, including the chance to review preliminary resource management alternatives and preliminary rationales for those alternatives. This early public involvement will help resolve conflicts and produce a Resource Management Plan that better reflects the needs of our citizens as well as others who use the public lands and have a stake in their future" (*Public Land News Bulletin*, 2016). The Public Lands Foundation (an association of retired BLM officials) and several conservation organizations likewise applaud the new planning rule for these and other reasons (*Public Lands News Bulletin*, 2016).

The BLM adopted the final rule reforming the planning process on December 12, 2016, toward the end of the Obama administration. Almost immediately, several groups, including the American Petroleum Institute, American Exploration and Mining Association, Independent Petroleum Association of America, Public Lands Council, and the National Association of Counties, opposed the rule, arguing, among other things, that it allowed "radical special interests" to have equal footing with local officials. These opponents called on Congress to repeal the administrative rule using the Congressional Review Act. During the first two months of the Trump administration, both the U.S. House of Representatives and U.S. Senate voted to revoke the rule, and President Trump signed the legislation on March 27, 2017. According to the Congressional Review Act, the BLM may not propose any new rule that is substantially the same as Planning 2.0. This means that BLM Resource Management Plans will be guided by antiquated planning rules that limit public participation.

These three examples—CFLRP, U.S. Forest Service 2012 Planning Rule, and BLM Planning 2.0—represent a step forward in the way public land management agencies engage citizens, stakeholders, and other governments. There is a significant difference, however, between this type of government-sponsored collaboration and (1) the experimental approaches explained above; and (2) the type of homegrown collaboration that has emerged organically in the COTC and throughout the American West.

Community-based collaboration, as well as the type of collaboration envisioned in the experimental portfolio, represents a fundamentally different type of decision making relative to the conventional model of expert decision making. Community-based collaboration is an inherently decentralized, democratic form of governing (Kemmis & McKinney, 2011). The experimental portfolio seeks to shift the locale of decision making from expert agencies to more of a shared decision-making approach. By contrast, government-sponsored collaboration is embedded within the expert model of decision making, a system and a culture that are inherently centralized and hierarchical. Community-based collaboration facilitates a shared ownership of the process, decisions, and outcomes. By contrast, government-sponsored collaboration is at best advisory, and thus resembles conventional approaches to public participation that "seek input and advice" but do not (and cannot) share decision making (Kemmis, 2001).[8]

Despite this fundamental difference, the innovative approaches federal land management agencies are using to foster public participation and shared problem solving represent a promising trend in public land management. Time will tell whether these innovations provide more direct and meaningful public participation, generate decisions that receive broad public support, and make implementation easier because the stakeholders have helped shape the proposed plans and programs.

Conclusion

Democracy is a work in progress, and any and all innovations and experiments to improve the process of public participation and shared problem solving in public land management are welcome. A diversity of approaches, bottom up and top down, is most likely to foster a healthy, high-functioning "ecology of governance." Highlighting the greater sage grouse conservation effort, Interior Secretary Sally Jewell recently argued that the future of public land management revolves around collaboration at the scale of large landscapes. "That's the model for the future of conservation. That big picture, roll-up-your-sleeves, get-input-from-all-stakeholders kind of planning is how land management agencies should orient themselves in the 21st century" (Jewell, 2016).

As calls for reform, experimentation, and innovation continue, it is important to acknowledge and respond to the legitimate issues and concerns that many people have raised since the emergence of the so-called collaboration movement (Golton, 1980; McCloskey, 1996; Kenny, 2000; Coggins, 2001; Dukes et al., 2001; Coulter et al., n.d.). Although a review of the arguments against collaboration is beyond the scope of this chapter, it is useful to emphasize that collaboration is not a panacea, that it does not replace existing environmental laws, and that agencies cannot abdicate their decision-making authority. As recently concluded by Martin Nie and Peter Metcalf (2015), both collaboration and litigation are necessary components of modern national forest management.

Despite the rhetoric about all the problems facing public lands in the American West, it is encouraging to see leaders from many walks of life searching for innovative approaches to address land, water, and related issues at different geographic and temporal scales; deal with complexity, uncertainty, and change; acknowledge and make sense of the diverse community of interests; and give citizens more meaningful opportunities to be involved in decision making. Collaboration is increasingly the forum of first resort for one simple reason—it works.

Notes

1. Portions of this chapter, including figures, were previously published in *Environmental Law Reporter* (2018). © 2018, Environmental Law Institute·, Washington, DC. Reprinted with permission from ELR·.

2. The Republican Party's natural resources agenda can be found at https://www.gop.com/platform/americas-natural-resources/. Among other things, it states that "Congress shall immediately pass universal legislation providing for a timely and orderly mechanism requiring the federal government to convey certain federally controlled public lands to states." By contrast, the Democratic Party Platform states that "Democrats believe in the conservation and collaborative stewardship of our shared natural heritage . . ." (p. 29): *2016 Democratic Party Platform* (July 21, 2016).

3. The ideological rationale underlying this sustained debate over "who" should own and manage public lands has been remarkably consistent: local people should have substantial authority over these lands because they use them regularly and understand them better than anyone else. Notwithstanding the legal, political, and economic arguments against the transfer of federal lands to states, this movement does not appear to be going away any time soon.

4. The following narrative draws heavily on Bates (2010). References to historical events, laws, and other details can be found in this policy report.

5. In addition to these three examples, it is instructive to review the degree to which negotiation, mediation, and collaboration has been integrated into agency decision-making and conflict resolution processes. For starters, see Van De Wetering & McKinney (2006), Copeland (2006), and U.S. Department of the Interior (2008). The administrative rule promulgated by the Secretary of the Interior in 2008 is particularly interesting in that it allows and encourages agencies and bureaus within the Department of the Interior to integrate consensus-based alternatives into their analysis as governed by NEPA. The administrative rule clarifies that there is no guarantee that any consensus-based alternative will be considered to be a reasonable alternative or be identified as the preferred alternative. Agencies and bureaus are required to explain how the consensus-based alternative is reflected in the proposed action and final decision. It's not clear how, if at all, this administrative rule has been implemented in practice.

6. The reference to "secret sauce" refers to a metaphor used by Hillary Tompkins, Solicitor, U.S. Department of the Interior, during her keynote address at

the University of Montana's 36th Public Land Law Conference (2015). According to the Solicitor, the first ingredient to successful collaboration is a catalyst, which often comes in the form of conflict. The second ingredient is public sentiment in your favor; you can't force the outcome. The third ingredient is the right messenger, someone people will listen to, trust, and respect. The fourth ingredient is the right setting, the place where solutions can emerge. The fifth ingredient is creative thinking, often out of the box.

7. See http://www.fs.usda.gov/main/planningrule/collaboration for a review of the collaborative process used to shape the 2012 planning rule.

8. Public land management agencies cannot delegate or devolve their Congressionally derived management authority to a collaborative group. When the National Park Service sought to delegate its authority for the Niobrara National Scenic River to a local council composed largely of local government officials and private landowners, the court concluded that the agency went beyond the scope of its authority to foster a cooperative approach to management (see *National Parks Conservation Association v. Stanton*, 54 F. Supp. 2d 7 (D.D.C. 1999). The governance arrangement was unlawful "because NPS retains no oversight over the Council, no final reviewing authority over the Council's actions or inaction, and the Council's dominant private local interests are likely to conflict with the national environmental interests that NPS is statutorily mandated to represent." The NPS retained only one seat on the eleven-member council, and the agency's only recourse if it was unhappy with the Council's decisions and direction was to terminate the cooperative agreement altogether. These same delegation principles seemed to limit the ability of the U.S. Fish and Wildlife to delegate its authority over endangered species management to a local Citizens Management Committee as proposed for grizzly bear reintroduction in the Selway-Bitterroot mountains. See also Van De Wetering, S. (2000).

References

Alexander, K., & Gorte, R. W. (2007). Federal land ownership: Constitutional authority and the history of acquisition, disposal, and retention. Washington, DC: Congressional Research Service.

Allen, C. R., & Garmestani, A. S. (Eds.). (2015). *Adaptive management of social-ecological systems.* New York: Springer.

Arnstein, S. R. (1969). A ladder of citizen participation. *Journal of the American Planning Association, 4,* 216–24.

Bates, S. (Ed.). (2010). *Remarkable beyond borders: People and landscapes in the Crown of the Continent.* Bozeman, MT: Sonoran Institute.

Bears Ears Tribal Coalition (2015, October 15). *The Tribal proposal to President Obama for the Bears Ears National Monument.* Retrieved from www.bears earscoalition.org

Blackfoot Challenge (2015). *2015 Annual Report*. Retrieved from www.blackfoot challenge.org

Bingham, L. B., Nabatchi, T., & O'Leary, R. (2015). The new governance: Practices and processes for stakeholder and citizen participation in the work of government. *Public Administration Review, 65*(5), 547–558.

Bixler, P., & Kittler, B. (2015, July). *Collaborative Forest Service landscape restoration: A meta-analysis of existing research on the CFLR program*. Washington, DC: Pinchot Institute for Conservation.

Brick, P., et al. (Eds.). (2001). *Across the great divide: Explorations in collaborative conversation and the American West*. Washington, DC: Island Press.

Brown, K. (2016, April 16). Bundy militia's takeover dreams dashed by bond between ranchers and Feds. *Courthouse News Service*. Retrieved from http://oldarchives.courthousenews.com/2016/04/06/bundy-militias-takeover-dreams-dashed-by-bond-between-ranchers-and-feds.htm

Center for Natural Resources and Environmental Policy (2015, February 19). *Public participation: Lessons learned implementing the 2012 U.S. Forest Service planning rule*. Bozeman, MT: University of Montana.

Center for the Rocky Mountain West and Institute for Environment and Natural Resources (2000, March). *Reclaiming NEPA's potential: Can collaborative processes improve environmental decision making?* Bozeman, MT: Center for the Rocky Mountain West.

Chaney, R. (2016, April 4). Blackfeet welcome bison herd to new reservation home. *Missoulian*. Retrieved from http://missoulian.com/news/state-and-regional/blackfeet-welcome-bison-herd-to-new-reservation-home/article_0a879d40-e473-5c99-aabc-c0c83b571790.html

Coggins, G. C. (2001). Of Californicators, Quislings and crazies: Some perils of devolved collaboration. In P. Brick, et al. (Eds.), *Across the great divide: Explorations in collaborative conversation and the American West* (pp. 163–171). Washington, DC: Island Press.

Conference of Western Attorneys General (2016, July 19). *Report of the Public Lands Subcommittee, Western Attorneys General Litigation Action Committee*.

Copeland, C. W. (2006). *Negotiated rulemaking*. Washington, DC: Congressional Research Service.

Coulter, K., et al. (n.d.). *Collective statement on collaborative group trends*. Unpublished manuscript.

Crown Managers Partnership (CMP). (2016). *Strategic conservation framework 2016–2020*. Retrieved from http://crownmanagers.org/strategic-conservation-framewo/

Dax, M. J. (2016, May 12). How Tribes led the fight over Badger-Two Medicine oil and gas leases. *High Country News*. Retrieved from http://www.hcn.org/articles/how-tribes-led-the-fight-over-badger-two-medicine-oil-and-gas-leases

Dukes, E. F., et al. (2001). *Collaboration: A guide for environmental advocates*. Charlottesville, VA: University of Virginia, The Wilderness Society, and the National Audubon Society.

Emerson, K., & Murchie, P. (2010). Collaborative governance and climate change: Opportunities for public administration. In R. O'Leary, D. M. Van Slyke, & S. Kim (Eds.), *The future of public administration around the world: The Minnowbrook perspective* (pp. 141–154). Washington, DC: Georgetown University Press.

Emerson, K., & Nabatchi, T. (2015). *Collaborative governance regimes.* Washington, DC: Georgetown University Press.

Forest Options Group (1998). *The Second Century Report: A report to the American people.* Concord, MA: Thoreau Institute. Retrieved from http://www.ti.org/2c.html

Golton, R. J. (1980). A "sellout" for conservation advocates or a bargain? *The Environmental Professional,* 62–66.

Gorte, R. W., et al. (2012). *Federal land ownership: Overview and data.* Washington, DC: Congressional Research Service.

Great Northern Landscape Conservation Cooperative (GNLCC). (2015, March). *Great Northern LCC activities and accomplishments 2014.* Retrieved from http://greatnorthernlcc.org/updates/gnlcc-2014-activities-and-accomplishments

Hays, S. P. (1959). *Conservation and the gospel of efficiency: The progressive conservation movement 1890–1920.* Cambridge: Harvard University Press.

Hirt, P. W. (1996). *A conspiracy of optimism: Management of national forests since World War II.* Lincoln: University of Nebraska Press.

Innes, J., & Booher, D. (2010). *Planning with complexity: An introduction to collaborative rationality.* London: Routledge.

Innii Initiative (n.d.). *Innii initiative: The return of the buffalo.* Retrieved from https://www.youtube.com/watch?v=6LJfPMoGMAg

Jewell, S. (2016, April 19). The next 100 years of American conservation. Retrieved from https://medium.com/@Interior/the-next-100-years-of-american-conservation-397c42b8f1f2

Keiter, R. B., & Ruple, J. (2015). *The transfer of public lands movement: Taking the public out of public lands.* Salt Lake City, UT: Stegner Center.

Kemmis, D. (2001). *This sovereign land: A new vision for governing the West.* Washington, DC: Island Press.

Kemmis, D., & McKinney, M. (2011). *Collaboration and the ecology of democracy.* Dayton, OH: Kettering Foundation.

Kenney, D. S. (2000). *Arguing about consensus: Examining the case against Western watershed initiatives and other collaborative groups active in natural resources management.* Boulder, CO: University of Colorado Natural Resources Law Center.

Klugin, T. (2016, April 10). Forging a new path: Regional trail users find common ground despite national controversy over bikes in wilderness. *Helena Independent Record.*

Leshy, J. D. (1980). Unraveling the Sagebrush Rebellion: Law, politics, and Federal lands. *UC Davis Law Review 14,* 317–355.

Locke, H. (2015, August 18). Pledge to restore wild buffalo unites First Nations of North America. *National Geographic*. Retrieved from http://voices.national geographic.org/2015/08/18/the-power-of-a-wild-animal-to-help-people-two-more-first-nations-sign-historic-treaty-to-bring-wild-buffalo-back-to-tribal-lands/

Marris, E. (2016, Summer). How Malheur became the epicenter of community-led conservation. *Audobon Magazine*. Retrieved from http://www.audubon.org/magazine/summer-2016/how-malheur-became-epicenter-community-led

McCloskey, M. (1996, May 13). The skeptic: Collaboration has its limits. *High Country News*. Retrieved from http://www.hcn.org/issues/59/1839

McKinney, M. (2016a, May). Personal conversation with Flathead National Forest planner.

McKinney, M. (2016b, September 11). Personal conversation with Emily Olson, Program Manager, National Forest Foundation.

McKinney, M., & Johnson, S. (2013). *Large landscape conservation in the Rocky Mountain West: An inventory and status report*. Bozeman, MT: Center for Natural Resources and Environmental Policy.

Mead, M. H. (2016, May 25). Letter from Governor Matthew H. Mead and Governor Steve Bullock to Ms. Leah Baker, Acting Branch Chair for Planning and NEPA, Bureau of Land Management.

Montana Governor Bullock announces workshops for Western Governors' National Forest and Rangeland Management Initiative (2016). Retrieved from http://westgov.org/news/montana-gov-bullock-announces-workshops-for-western-governors-national-for

National Parks Conservation Association v. Stanton, 54 F. Supp. 2d 7 (D.D.C. 1999).

Nelson, R. H. (1995). *Public lands and private rights: The failure of scientific management*. Lanham, MD: Rowman and Littlefield.

Nelson, R. H. (2015). *Charter forests: A new management approach for national forests*. PERC Policy Series 53. Bozeman MT: Property and Environment Research Center.

Nie, M., & Metcalf, P. (2015, October). *The contested use of collaboration and litigation in National Forest management*. Bozeman, MT: Bolle Center for People and Forests, University of Montana.

Omnibus Public Land Management Act of 2009 (2009, March 30). Public Law 111–11.

Ostrom, E. (2009, December 8). *Beyond markets and states: Polycentric governance of complex economic systems*. Nobel Prize lecture. Retrieved from www.nobelprize.org/nobel_prizes/economic-sciences/laureates/.../ostrom_lecture.pdf

Public Lands News Bulletin (2016, July 11). Bulletin #7, 1–3.

Public Policy Research Institute (2006). *The legal framework for cooperative conservation*. Bozeman, MT: University of Montana.

Rasker, R. (2012). *West is best: How public lands in the West offer a comparative economic advantage.* Bozeman, MT: Headwaters Economics.

Schultz, C., & Nie, M. (2012). Decision-making triggers, adaptive management, and natural resources law and planning. *Natural Resources Journal, 52,* 443–521.

Silken, J. R. (2009). *The nation's largest landowner: The Bureau of Land Management in the American West.* Lawrence: University Press of Kansas.

Snow, D. (2001). Coming home: An introduction to collaborative conservation. In P. Brick, et al. (Eds.), *Across the great divide: Explorations in collaborative conversation and the American West* (pp. 1–11). Washington, DC: Island Press.

Snow, D., et al. (1998). The Lubrecht conversations: A diverse group discusses the future of the Forest Service in a changing West. *Chronicle of Community, 3*(1), 5–16. Reprinted in P. Brick, et al. (Eds.), *Across the great divide: Explorations in collaborative conversation and the American West.* Washington, DC: Island Press.

Stegner, W. (1969). *The sound of mountain water.* Garden, City NY: Doubleday.

Tompkins, H. (2015). Transcending boundaries: Achieving success in cooperative management of natural resources. Keynote Address, 36th Public Land Conference, University of Montana.

U.S. Democratic Party (2016). *2016 Democratic Party platform.* Retrieved from www.presidency.ucsb.edu/papers_pdf/117717.pdf

U.S. Department of the Interior (2008). *Incorporating consensus-based management.* 43 CFR 46.110.

U.S. Department of the Interior (2016a). National fish, wildlife, and plants adaptation strategy. 2016 Leadership Award recipients. Retrieved from https://www.wildlifeadaptationstrategy.gov/award-recipients.php

U.S. Department of the Interior (2016b, November 22). *Resource management planning.* Bureau of Land Management 43 CFR Part 1600.

U.S. Forest Service (2012, April 9). *National Forest System land management planning.* 36 CFR, Part 219.

U.S. Forest Service (2015a, April). *Collaborative Forest Landscape Restoration program 5-year report, FY 2010–14.* FS-1047. Retrieved from https://www.fs.fed.us/restoration/documents/cflrp/CFLRP_5-YearReport.pdf

U.S. Forest Service (2015b, January 30). *Forest Manual 1900-Planning.* Chapter 1920—Land management planning. Retrieved from https://www.fs.fed.us/im/directives/fsm/1900/wo_1920.doc

U.S. Forest Service (2016, June). *A citizens' guide to national forest planning.* Prepared by Federal Advisory Committee on Implementation of the 2012 Land Management Planning Rule. Retrieved from https://www.fs.usda.gov/Internet/FSE_DOCUMENTS/fseprd509144.pdf

U.S. House of Representatives (2016, July 7). Statement of Jeffrey Fontaine, Executive Director, National Association of Counties, before the Subcommittee on Oversight and Investigations Committee for National Resources, U.S.

House of Representatives Hearing on State Perspectives on BLM 's Draft Planning 2.0 Rule.

U.S. Republican Party (2016). *2016 Republican Party platform*. Retrieved from https://www.gop.com/platform/

Van de Wetering, S. B. (2000). Bitterroot grizzly bear reintroduction: Management by citizen committee? In P. Brick, et al. (Eds.), *Across the great divide: Explorations in collaborative conversation and the American West* (pp. 150–159). Washington, DC: Island Press.

Van de Wetering, S. B., & McKinney, M. (2006). The role of mandatory dispute resolution in federal environmental law: Lessons from the Clean Air Act. *Journal of Environmental Law and Litigation, 21*, 1–45.

Weaver, J. L. (2015, May). *Sacred lands: Innovative conservation of wildlife and cultural values Badger-Two Medicine area, Montana*. Wildlife Conservation Society Working Paper. Retrieved from https://library.wcs.org/doi/ctl/view/mid/33065/pubid/DMX2867800000.aspx

Weber, E. (2003). *Bringing society back in: Grassroots ecosystem management, accountability, and sustainable communities*. Cambridge, MA: MIT Press.

White House (2016, December 28). Presidential proclamation—Establishment of the Bears Ears National Monument. Retrieved from https://obamawhitehouse.archives.gov/the-press-office/2016/12/28/statement-president-designation-bears-ears-national-monument-and-gold

Whitefish Range Partnership (2013, November 18). *Whitefish Range partnership agreement*. Retrieved from headwatersmontana.org/sites/default/files/WRP_Final_11_18_2013.pdf

Wild Montana (n.d.). Our work-protecting public lands. Retrieved from http://wildmontana.org/our-work/protecting-public-lands/quiet-trails/montana-high-divide-trails/

Wondolleck, J. (1988). *Public lands conflict and resolution: Managing National Forest disputes*. New York: Springer.

Yellowstone to Yukon Conservation Initiative (Y2Y). (2016). *2015 annual report*. Retrieved from http://2015report.y2y.net

Chapter 4

Speaking of Place

Analysis of Place-Based Discourse in Participatory Decision Making

COLENE J. LIND

M any literatures suggest that people lose something important when they forget where they are. Scholars in fields as diverse as journalism (Hess & Waller, 2014; Thomson, Bennett, Johnston, & Mason, 2015), marketing (Brocato, Baker, & Voorhees, 2015), politics and public administration (Nabatchi & Amsler, 2014; Walsh, 2012), information sciences (Seggern, Merrill, & Zhu, 2010), art (Ippolito, 2009), history (Cocciolo & Rabina, 2013), and education (Brooke, 2012) put place at the center of decision making. In environmental communication, place-based discourse is said to ground problems in local ecologies and encourage inclusive decision making that is more likely to engender stakeholder support (Cantrill, 2012).

But at the same time, the rhetoric of place has been linked to dele-terious outcomes in a wide range of human conflicts, from NIMBYism to militant nationalism and genocide (e.g., Devine-Wright, 2009). Based on these findings, should participatory processes put locales and local citizens at the center of public decision making about the environment, or should conveners be more concerned with transcending the narrow interests of particular communities? Providing guidance on this question demands a better understanding of how participants purposively use the language of place in public discussions about the environment.

Toward this end, I conducted rhetorical analysis to describe how members of the public, invited to express their opinions on water-use policy, talked about place. After reviewing current thinking about place-based discourse in environmental conflicts, I share findings from my analysis of transcripts from 30 meetings. Overall, I found that through the language of place, speakers effectively contributed to the conversations, but they also failed to account for the effects of human communities on each other. I conclude by considering how participatory processes might be shifted and studied further to encourage the full potential of place-based discourse.

The Paradox of Place

According to Cantrill (2012), "[t]he centrality of place-based discourses in the process of planning and policymaking . . . has become pinned to academic conventional wisdom" (p. 10). This wisdom proceeds from the connection of political power to proxemics. Kemmis (1990, p. 18) begins by positing that "no real [political] culture can exist in abstraction from place yet that abstraction is one of the hallmarks of our time." He laments that time has come to dominate space in modern politics (Hart, Childers, & Lind, 2013), with political power residing in opportune and fleeting coalitions among interests. For while "Montesquieu wanted people face to face and in touch with each other, Madison wanted them dispersed, disconnected, out of touch . . . precisely in order to keep people from too much beholding of a 'common unity'" (Kemmis, 1990, p. 18). To combat the disempowerment of division, people need place to find common ground, literally and figuratively. As Wendell Berry (2003, p. 135) suggests, "people who live at the lower ends of watershed cannot be isolationists," for "pondering such facts of gravity and the fluidity of water shows us that the golden rule speaks to a condition of absolute interdependency."

If public cannot exist without place, then neither can human life. Kemmis (1990, p. 117) reminds that "there are not many rivers, one for each of us, but only this one river, and if we all want to stay here, in some kind of relation to the river, then we have to learn, somehow, to live together." Not just ecological features but built environments, too, demand that those who wish to dwell in the same space learn to share it. Be it an endless stretch of empty beach or a cozy cafe, humans develop emotional bonds with places, imbuing them with special meanings (Devine-Wright,

2009; Milligan, 1998). Such place attachments usually become apparent only when disrupted, as when the beach seems crowded or the cafe unexpectedly closed (Milligan, 1998). But at these times we recognize what is always true: the finitude of place makes it necessary to collaborate with others to continue enjoying its benefits.

Consequently, place-based discourses signal and reinforce cognitive connections between speakers and their surroundings. Indeed, as Cantrill and Senecah (2001, p. 186) claim, it is the "cognitive representations of the self-in-place" that "form the major link between how people appraise communication appeals and how they behave in the environment." Empirical studies support this association. Cantrill and colleagues (2007), for example, found that homeowners referenced place-based considerations (i.e., "I can walk to the store," "there are deer here") when explaining where they chose to live, be it the urban center or suburbia. Similarly, Druschke (2013) found that Iowa farmers who symbolically identified with a watershed drew on their sense of place as motivation for implementing conservation practices. A place-based discourse, therefore, draws on and allows participants to deepen their place attachments, encouraging more environmentally benign actions.

And because places are spaces that have been infused with human meaning, they serve as repositories of material and symbolic interactions. In support of this point, Procter (1996, p. 223) contends that the idea of a place comes from "rhetorical communities communicating different values, social relationships, and actions in response to the same landscape." Or, as Carbaugh and Cerulli (2013, p. 7) put it, the "place-setting function of communication is . . . deeply and radically cultural." Thus, a group with a common understanding of a place also shares symbolic resources. Because environmental issues involve "complex and contested concepts that invoke value judgments based on different beliefs," a viable community response necessitates the convergence of "like-minded people who share similar belief systems that make sense of the world and their place within it." Toward this end, place-based discourse "creates the conditions for collective action to address the opportunities and challenges associated with sustainability" (Ward & van Vuuren, 2013, p. 64).

Finally, place-based discourses invite participants to take responsibility for policy making, rather than ceding decisions to elected representatives, agency officials, or technical experts. Writing of Forest Service land-use decisions, Daniels and Walker (2001, p. 5) report that, traditionally, "the agency finds itself operating as an arbiter in an environment where each party faces incentives that encourage extreme behavior and the worst sort of positional rhetoric, but faces no incentive to learn from or accommodate one another." In contrast, place-based processes shift the locus of control from officials to stakeholders. As a result, "once the parties themselves get the ideas that they are responsible for coming up with the answer, rather than simply turning it over to a third party, they are very likely to begin to think and believe differently" (Kemmis, 1990, p. 113). The process not only sounds different—more collaborative than confrontational—but it also begets greater trust, less social isolation, and more widely shared commitment to the decision (Daniels & Walker, 2001; Kemmis, 1990).

For these reasons, environmental sustainability surely is facilitated, at least in part, by the language of place. However, place-based processes and their discourses have critics, too. To begin, locally controlled planning potentially contravenes regional and national goals, with citizen engagement becoming "a form of permanent nimbyism that can easily degenerate into pork-barrel politics" (Coleman & Firmstone, 2014, p. 842), particularly in communities with high levels of social capital (Peterson, Peterson, Peterson, Allison, & Gore, 2006; see also Zarlengo, 2016). Similarly, economic interests have been known to effectively resist environmentalism by claiming to best represent indigenous concerns (e.g., Shellabarger, Peterson, Sills, & Cubbage, 2012). Certainly local opposition often stems not from social privilege but from years of "environmental injustice in the processes for site selection and local participation in decision making" (Endres, 2012, p. 329). Even so, the larger dilemma remains: "Effective bridging between local contention and geographically wider struggles is rarely straightforward or forthcoming, evidenced by the relative paucity of examples in the environmental politics" (Usher, 2013, p. 813).

Talk of place might not only contravene large-scale coordinated action but also devolve into antagonistic, identity-based conflict as well. As Kenneth Burke (1950, p. 23) presciently observed, unity implies division with something or someone. Consequently, "the idea of being a member of a local community may come to the fore when people face the presence of . . . those who are considered outsiders" (Marzorati, 2013, p. 254). Illustrating this phenomenon, Marzorati found that long-time residents of an Italian village discursively constructed a public park "as a parochial realm that the group of residents considered more entitled to use than people who come from somewhere else," namely recent immigrants. Through her interviews with the residents, Marzorati demonstrates that the "existence of a local community, intended as a spatialised sense of 'us' . . . can be rhetorically evoked under specific circumstances" to unite *and* divide (p. 254).

As Marzorati's findings indicate, place-based discourses also lean toward provincial perspectives that eschew change in favor of the status quo. For example, based on his interviews with working-class citizens, Robert Lane (1962) concluded that those with community-based identities tended to be more immobile in their thinking. By contrast, those with a cosmopolitan view pluralized the world in a way that provincials could not. Similarly, cultural myths can leave little room for progress in the rural communities often at the center of natural-resource controversies. Ron Lee (1995) investigates one such myth equating place and virtue. This master narrative presents small towns and their people as simple, genuine, pre-political, and insulated from the nefarious machinations and cultural divisions of the urban scene. Ill-advised as were her comments, Sarah Palin's expression of affection for the places where "real Americans" live indicates the continuing persuasiveness of such beliefs about rural America (Rice, 2012). Consequently, Lee (1995, p. 19) wonders "how political rhetors can transcend the American 'geographic mentality' and build moral arguments on something other than the homogenous myths of the American landscape." He argues that Jimmy Carter and Bill Clinton tried connecting the small-town mythos to diverse opinion and innovation, but each failed in his own way.

Whether set in rural or urban locations, planning processes that privilege place-based discourses have been said to sacrifice quality for consensus. Peterson and colleagues (2006, p. 136) remind that in the United States alone, community-based planning has resulted in the "National Park Service eradicating predators in parks, irrigators cheating the government out of reclamation subsidies, and grazing advisory councils dictating winners and losers in federal forage allocation to the detriment of small ranchers and nomadic sheepherders." Perhaps this is because consensual processes give "a major advantage to the participants most resistant to restrictions," and why the World Wildlife Fund, for example, recommends consensus "only in very limited circumstances" (Center for International Environmental Law, 1999, p. iv).

Finally, without institutional checks and balances, place-based discourses allow powerful interests to dominate policy making. Through contrasting case studies, Peterson and colleagues (Peterson et al., 2006) demonstrate that the social capital of local assemblies "must be shared across polarized views and across hegemonic configurations" (p. 135); otherwise, place-based processes "lead to exclusion of outsiders, encourage free-riders, and may lead to the restriction of personal freedoms" (p. 119). Similarly, Usher (2013) found that a local environmental group was only able to compete rhetorically when it adopted the hegemonic discourse of science and technology. Usher concludes that "indeed, it would certainly appear that money can buy knowledge and knowledge can in turn draw power," leading some to wonder if participatory process, no matter the scale, can escape the injustices of existing social systems (p. 823).

In sum, the ideals of place-based discourses are potentially undone by the realities of language and its use. On one hand, speaking of place invites communities to address local problems on their own terms while also recognizing and building human community and richer relationship with the environment. But rhetorically, reference to place is potentially as exclusionary as it is welcoming. Can public discussion that embraces place further decision making on the environment without presuming existing social, political, and economic arrangements? Also, can place-based discourses engage community members in public processes of local determination while also considering the broader impacts of policy?

Context and Method: Local Considerations of Water Use

To address these questions, I sought to learn more about place-based discourse as deployed in a series of community conversations about water supply. Around the globe but especially in the western United States, water presents itself as the most immediate and critical natural-resources issue (Sanderson & Frey, 2014). In western Kansas, for example, agriculture, oil and gas mining, and households rely on fossil water from the Ogallala Aquifer. Water has been pumped from this underground reservoir at rates that exceed recharge, with some sections already depleted past the point of viable extraction. The Kansas Geological Service estimates that most of the aquifer has a usable lifetime of fewer than 50 years, with broad sections of the deposit to be functionally gone in fewer than 20. Central and eastern Kansans face water problems, too. Here, state and federal reservoirs supply water, but holding capacity has been significantly compromised by siltation from soil erosion, reducing the storage of some constructed lakes to less than 50 percent of their original design. All parts of the state face water-quality challenges, particularly in the southeast, due to mining, and in the central region due to atrazine and nitrogen contamination.

With these problems in mind, in October 2013, the administration of Governor Sam Brownback initiated a statewide process to develop a 50-year water plan. The process included a series of public meetings to gather input for drafting the document. As a final step in crafting the plan, in January 2015, the Kansas Water Authority (KWA) convened 14 regional teams across the state, charging local groups of unelected citizen-stakeholders with drafting three to five water-use objectives for the respective regions. Consisting of citizens with local expertise and interests, these teams had great latitude. All of them were encouraged to solicit public input, but they were also directed to consider such information as advisory, with team members ultimately responsible for proposing local water-use goals to the KWA. Most regional teams hosted two or three public meetings to solicit community input.

These open sessions offered an opportunity to study place-based discourse in a public, participatory process of environmental decision making. From February through April 2015, I attended eight meetings hosted by the advisory teams, each held in a different Kansas town.[1] All meetings opened with a welcome from a state official and a technical debriefing from a state staffer. The welcome messages emphasized the administration's desire to hear from "average Kansans," under the assumption that "Kansans have good ideas."[2] The content of the technical updates varied at each meeting, addressing locally relevant conditions. After the welcome and briefing, trained facilitators from the Kansas State Cooperation Extension Service and the Kansas State University's Institute for Civic Discourse and Democracy led a nominal group process, asking participants to work in small groups.[3]

Each set of eight to ten participants discussed four questions about water supply in their region. Conveners at all sites included the following query: "What role should water conservation and public education play in meeting future needs?" The remaining three questions at each site invited talk about a topic salient in the area. For example, in communities where land use mainly consisted of row-crop farming, participants were asked how they "view water quality and nutrients in connection with water supply." By contrast, only participants in Western Kansas considered "how can the state and local management encourage economic growth along with conservation?" Similarly, only attendees in Hiawatha talked about what they would "most like to see protected should a transfer be pursued," a reference to a proposed cross-state canal to take water from the Missouri River to Western Kansas. Conveners at only one meeting site explicitly asked about economics, while two forums asked for ideas about a specific place.

Discussions of the four posed questions lasted about 50 minutes. I recorded 32 of these small-group dialogues. Transcripts from 30 of the recordings served as the texts for this study. My analysis presumes that these texts are the artifacts of rhetorical communication, undertaken to influence particular audiences within a certain context. In this case, the immediate audience included other citizens attending the meeting as well as the public officials who would review meeting notes and implement water policy. And because rhetoric is social action, it can be assessed based on practical and moral consequences (Tracy, 2001, p. 243). Here, I focus on the practical consequences for environmental decision making, seeking to discover how the patterns of symbol use could encourage or inhibit consideration of some topics or perspectives. After reading each transcript once, then highlighting references to specific locations, I repeatedly reviewed and compared talk about places throughout the 30 dialogues. I considered the broader context of the discussion whenever place was mentioned, looking for patterns of content and style and considering the rhetorical function of the place-specific mentions.

Place and People in Public Discussions About Water

Starting with their self-introductions, which almost invariably indicated from where each group member hailed, place figured prominently in the talk examined here. Most obviously, speakers referenced place to exemplify whatever point they were trying to make, allowing participants to contribute personal insights to the collective conversation. Without mention of place, most participants would seemingly have had little to say about water and its use. Closer reading, however, revealed several additional purposes served.

HERE

From telegraphic references ("I'm from Tuttle so I have a couple of specific Tuttle type of things") to lengthy descriptions, the transcripts were replete with talk about local places. In the east, watersheds often served to locate participants, such as "I live in the Kansas River basin but we have a business building that is actually in Missouri. So kind of the two watersheds." At the other end of the state, participants mostly mentioned their agricultural practices, as did this farmer, speaking in Goodland: "Our farm, at five quarters of ground, three wells, and we're finished [with irrigation]. So we're back to just wheat only." With access to few rivers and fewer reliable groundwater sources, Central Kansans turned to other placed-based affiliations to show their colors: "I'm in northern Lincoln County, and I've went strictly no-tilling and we don't have near the runoff we used to have. It's changed the way water flows." All three of these participants referred to some land-based feature to socially identify themselves, demonstrating that places define people as much as people create places.

Whether their stories were positive, as was the farmer's from Lincoln County, or ambivalent, like that of the landowner in Goodland, descriptions of local conditions also served as good reasons for listeners to accept a speaker's assertion (Fisher, 1987). As they gave concrete details of reservoirs rising or falling, stream banks shifting, or wells being drilled or going dry, participants contributed meaningful content in a style that signaled their conviction (Polletta & Lee, 2006). A participant in Manhattan, for example, spoke of the federal reservoir five miles away: "All you have to do is drive to the north end and it's ridiculous what's going on," he said, referring to stream-bank erosion. The tone of such totalizing, incredulous statements ("all you have to do," "it's ridiculous") contrasts with the more uncertain tenor of time, occasionally voiced in the meetings, too, as heard from this participant from Manhattan:

> The problem with reservoirs, they were built to fill in. And they had a life, maybe 50–60 years some maybe 100 years but the intent was to fill up and then build another reservoir somewhere. Unfortunately what costs millions back then now cost billions and people don't want to give up their lands for the reservoir so that makes it even more restricting but I think it's going to take conservation and it's going to impact agriculture because the domestic and municipal use and the electric power out of the three plants and including all the people within a district . . . about 1.1 million people in the state so that's a lot of people and you have to have the water for them to live on. . . . We're going to have to have some sort of change of crops to extend it because you're not going to be able to stop the sedimentation.

Using qualifiers, hedges, and quantifications to delimit his claims, this participant seemed less confident than his very sure neighbor. Both residents spoke of the same place, but the second statement illustrates that the vagaries of time stylistically diverge from the confidence of solid ground.

Indeed, the language of place seems tangible even when unseen systems are invoked. The following participant from Salina, for example, turned to the local to explain the otherwise complex interaction of water quality and quantity. Simultaneously, he argued for keeping water pure:

> So the city of Salina . . . they explored this but the cost of the pipeline was worse than the cost of treating the water, was so much so that it really took [the pipeline] out of the equation. So the worse we degrade the water, the more expensive it becomes for anyone to use it for water supply.

Occasionally, speakers at these meetings referenced local sites to introduce uncertainty into the conversation, as did the first speaker quoted below, at a Wichita-area meeting:

> SPEAKER ONE: By June of 2013 we all saw what Cheney Lake looked like and whether or not that was going to be able to supply close to half a million population, and what were the Equus beds, what was the level doing in the Equus beds during the same time? And I know Wichita moved towards taking more water out of Cheney and less out of the Equus Beds during that time period, isn't that right?

> SPEAKER TWO: Yeah it really, there was a shift during the drought and with more reliance, ultimately, as Cheney started to go down there was a shift in trying to take more of the demand out of the Equus Beds during that period to try to conserve Cheney as long as possible.

But even as Speaker One wondered aloud about a place in the past, he sought to gain clarity rather than to raise doubts.

Overall, the discourse studied here affirms the notion that talking about places allows diverse participants to contribute to environmental discussions meaningfully and productively. The experiential knowledge of the haunts we know firsthand might limit generalizability of the information offered, but nevertheless, rational public decision making does not require that all participants hold equal knowledge. Rather, speakers need only assert what they believe in a sincere and cooperative search for truth (Habermas, 1984). The boots-on-the-ground perspective empowers experts and non-experts alike to so do on equal footing.

THERE

While local ground dominated talk in these public conversations, more distant features figured in the discussions, too. In some instances, participants lifted up distant places as an optimistic ideal, as did a Hiawatha participant who recounted seeing "a very interesting Discovery Channel [program] on a goat herder from Sudan who had taken the arid desert and, through working with natural animals that worked into burrowing in the desert, he now has reclaimed the area and has a forest." Other places were offered as cautionary tales, warning of what might happen here. For example, two participants in Goodland referenced the economic decline and depopulation of a Colorado town 75 miles away:

> SPEAKER ONE: I sat on a zero-depletion committee 20 years ago and we sat around a table like this in Colby for two hours trying to decide economics of the impact of zero depletion, and I said, have we got a couple hours left in the day? Let's get in a van and drive to Flagler and look around and then we can come back. And that's what it's going to look like and you don't have to worry about it anymore.

> SPEAKER TWO: That's what it's going to look like in 50 years if we keep going at this rate?

> SPEAKER ONE: Right.

Notice that both of the speakers invoked other people as they spoke of other places. The goatherd gets a direct mention while people in a neighboring town are noted, obliquely, for their absence. In the same way, a speaker in Wichita argued for changing current laws disallowing the reuse of municipal water: "In San Diego, California, they use direct reuse; they do in Orange County, California. Wichita follows Texas and there's many states that are looking at that." In all three examples, the actions of people in other places—not the places themselves—serve a didactic function in the discourse.

In these examples and others, it is the combination of people and place that makes distant locales germane. In the following extended discussion from the Ottawa meeting, participants compared notes on several different places and in the process constructed a common, nationwide threat to water supply and the need for conservation:

> PARTICIPANT ONE: I mean let's face it, it's a concern all over the U.S., and I've got a son that lives in Dallas so they've been under water restriction for six or seven years—
>
> PARTICIPANT TWO: I have a brother in Wichita Falls and he's considered—
>
> PARTICIPANT ONE: —yeah and they do way more of this water conservation and educate the public than we've ever thought about around here.
>
> PARTICIPANT THREE: Does Wichita Falls, is that where they're converting the actual from the wastewater feeder plant?
>
> PARTICIPANT TWO: I think they're taking from the wastewater plant and . . . blending it, and it's easier to treat than the surface water, and it's cleaner than surface water anyways.
>
> PARTICIPANT ONE: Yeah like where my kids live, everybody around there, I mean, doesn't irrigate. Now they have different rules in the state of Kansas because what they were doing wouldn't be legal in Kansas, but I mean they've been doing that when they first moved down there out of college and there were 3,200 apartments in this complex, and they were using all the [gray] water for their grass and golf courses and everything else was wastewater.

Perhaps "the son that lives in Dallas" and the "brother in Wichita Falls" gave the respective participants permission to offer information about distant places—information they otherwise could not directly assess and, therefore, could not credibly present. Human relations, therefore, give talk of distant places an authority similar to that of local experiences.

But because removed places also connote other peoples, the people-place nexus permits participants to speak of others without mentioning them by name. Using this linguistic move, a participant in Goodland offered an enigmatic proverb, saying, "sometimes people in the state just don't get it. I was always told we have more water problems because it's downhill to Topeka and uphill to Denver." What the Western Kansan intends to convey about water is unclear. And yet the speaker plainly communicates his antipathies toward urbanites in Denver and state officials in Topeka without mentioning either. At the same time, the speaker demonstrates the power of language to invoke a sense of communion among one group by promoting division with another (Burke, 1950).

HERE VERSUS THERE

Participants in local discussions also fostered common understanding through place-based comparisons. For example, when asked by a facilitator about confined cattle feeding in the area, an Ottawa participant instinctually turned to place, answering, "not like the big mega ones out west but there's a few thousand head." In addition to such mundane habits, comparing places empowered participants to do at least three more things.

First, speakers established that ecologies are different from one area to the next. For example, a Wichita participant observed the following:

> There's varying conditions across the Equus Beds in regard to recharge rates. You know, Sedgwick County, Harvey County, Reno County all have relatively high recharge rates compared to what McPherson County has, so you have kind of different conditions across the area, too, so it's not necessarily all one size fits all for what the solution would be.

As the speaker hints ("so it's not necessarily one-size fits all"), differences in local conditions also suggest the need for flexible policy responses. Similarly, other citizens argued that differing environmental conditions justified differing personal actions, depending on where one lived. Note that in the following exchange between speakers, the interlocutors conclude that communities with more rain need less education and, by extension, should be less conservative with water.

> SPEAKER ONE: When you're talking about conservation, the shoe doesn't fit all. When you're talking about western Kansas, and

here it's not the same. And we're talking about education, that's where you probably need to educate people. It's not the same in this region as it is out there.

SPEAKER TWO: Even just look on that map. Trego County's usage compared to Brooks County—it's a lot different.

SPEAKER THREE: Depends on how much you get from the sky. The less you get from the sky, the more conservative you are. Automatically.

Speaker Three may or may not be correct. Either way, the trio expressed a reasonable, ecologically sensitive position: people should behave differently in different locations, in concert with varying natural conditions. Interestingly, however, the discussants used their place-based argument to resist conservation education in their community—a position that is hard to square with environmental sustainability. And Beloit participants were not alone; at least one person at every meeting analyzed here argued that conservation efforts should be driven by local supply. As one Wichita speaker pithily summarized, "some areas have plenty of water," so why waste the effort?

Second, speakers compared the severity of their water problems to conditions in other places. In doing so, they assured listeners that things were bad here but worse elsewhere. Sometimes, comparisons offered good reason for locals to take preventive action to avoid the calamities befalling others. One Wichita resident observed that she and her neighbors were "in such a better position [than] in some places in the state"; consequently, "we really have to protect that condition." In calling for action on soil erosion, a Manhattan participant admitted that "we're not worried about Ogallala here," but nevertheless, the local reservoir "is getting smaller too and harder to get to." Conversely, relative advantage also justified inaction, as the following exchange demonstrates:

SPEAKER ONE: My point would be the conservation level here shouldn't be the same throughout the other areas.

SPEAKER TWO: Define conservation.

SPEAKER ONE: . . . This Solomon Valley and Republican water basin does not have the water-supply issues of the western part of the state, and it doesn't have the reservoir siltation problems of the eastern part of the state. So we should be encouraging to be cautious of the initial goal-setting document. The initial vision goal is being set at 20 percent reduction across the state. That does no good for this area. 20 percent is not enough in some parts of the state, but it would be punishing those in this area.

Notably, there are no examples in the transcripts of citizens using such comparisons to express solidarity with or sympathy for people in other locations. Rather, participants tended to distance themselves from others, especially farmers in western Kansas. To wit, one Salina participant said explicitly what many seemed to be thinking: "We've got to stop raising corn and alfalfa in western Kansas."

Third, as speakers compared places, they also voiced what people will gain or lose in light of changing environmental conditions. Usually, such talk was framed as loss, with upstream actions negatively influencing local conditions (Tversky & Kahneman, 1981). Occasionally, however, speakers adopted a gains frame, presenting the environmental losses in one region as an economic opportunity in another. For example, the depletion of the Ogallala Aquafer is often equated with environmental catastrophe (Sanderson & Frey, 2014), but the following Wichita participant argues from a glass-half-full perspective:

My comments are that we need to use it up to the sustainability of the Equus Beds because it is fresh water underground source for us and with the Ogallala [Aquafer] drying up the economic liability can move east or for us and be closer to Wichita to drive some economic value into Wichita, and also the export market that we conserve is great for the benefits of the state so I think the aquifer they just need to be careful with it I don't think it's being overused at this point.

Taken together, the three strategies—unique ecologies justifying local control, comparison, and relative gain or loss—suggest that the language of place holds promise and pitfalls for inclusive environmental decision making. When using these strategies, participants spoke of what made a locale unique. We might cheer such a discursive orientation, which ought to encourage meaningful consideration of the interaction of humans with non-human elements of the environment. But from this analysis, it is also clear that place-based talk does not necessarily coincide with holistic consideration of places.

LIMITS OF PLACE

This analysis has so far demonstrated that talk of place coincided with other kinds of reflections in patterned ways. However, I also found a conspicuous absence in the dialogues. Specifically, my reading produced no evidence that the language of place overlapped with consideration of community connections.

When reviewing references to places, I found examples of participants naming the local consequences of upstream actions. In North Central and Northeastern Kansas, for instance, speakers referenced agricultural practices high in the watershed and the harmful effects on their water supply. Similarly, residents in Southeast Kansas connected past mining activities with current water contamination. But while they mentioned place, never in the transcripts do any of the participants indicate how local actions might impact places *downstream*.

To my point, several groups wondered about alleviating shortages by holding more water in federal reservoirs for irrigation and domestic purposes. One Manhattan participant recognized this as a hard sell because such lakes "were typically built for flood control, so if you raise the conservation up a foot they're basically giving up a foot of flood storage." This comment shines a light on the trade-offs of a policy action and specifically identifies the federal agency with a mandate contrary to local interests. As telling, the participant makes no mention of the downstream communities that would face greater flooding risks if more water were held, nor do any other participants in 30 cumulative hours of discussion.

Based on what they said, participants recognized that people are connected, with local actions having implications elsewhere. A participant from Hiawatha framed it thusly:

> There are a couple of levels, you have the personal level, then the community level, then the state, regional and national and global. I think there has to be some sense of a person at that smaller level understanding how that affects all those other levels too and how the governance happens at each one of those levels.

In her comments, however, this speaker does not reference any particular places. Meanwhile, those who spoke of human connections in tandem with specific locales assumed a rights-based perspective. A Goodland participant, for example, highlighted the political power of "Shawnee County or Johnson County or Wichita" as being "more important to our survival because" people in these places "control the state legislature. If they vote, it doesn't matter what the rest of Kansas wants." Certainly this speaker sees his community's fate as tied to others, but the connection is mechanistic and one-way rather than holistic and reciprocal. Similarly, a Salina participant connected different places based on their shared plight, indicating the following:

> The city of Des Moines had to put in $25 million worth of expense just to deal with the nitrates and a lot of little towns around here are greeted with a piece of paper that if you are a pregnant woman you should not be drinking this water and it seems to me this is the United States of America for God's sakes we shouldn't have to put up with that.

But even this statement connects places through national identity and the rights therein entailed—not based on a common reliance on the industrialized food system and the need for clean water or as a part of a web of actions and outcomes.

Indeed, the discourse of place in these meetings seemingly pushed out talk of shared obligations and outcomes. Over a few hundred pages of transcripts, I found exactly one example like the following exchange, heard in Salina:

> SPEAKER ONE: The good news is, just imagine how much more difficult this conversation becomes 90 miles west of here and then 180 miles west of here because you're in an aquifer. [We're] in an alluvial that recharges very well, your Kanopolis Lake does fill so we're having a great discussion and input but do you have people west of here . . .

> SPEAKER TWO: But what about the people downstream of us?

> SPEAKER ONE: From us?

> SPEAKER TWO: Yeah. If we don't conserve so that a lot of this water that we're blessed with goes downstream it just shows the problem.

To Speaker One, who emulated several of the rhetorical tactics of place-based discourse observed elsewhere in these transcripts, Speaker Two's question surprised and perhaps even seemed impertinent. To Speaker One, "people downstream" were off topic, for this was a forum about locals and their concerns. At best, "people out west"—upstream—might be worth discussing. The complete lack of consideration for those at lower elevations leaves us to wonder, is it possible to design public processes that invite more questions like those heard from Speaker Two?

Modifying Public Practices in Consideration of Place

From this analysis of how place functions in public discussions of environmental policy, we can conclude that it might help but also hinder. Talk of place meaningfully engages participants, and it also allows for considerations of local issues. Both facets make place-based talk a constructive resource for community-driven, consensual processes. In addition, this analysis demonstrates that talk of place does not necessarily limit participants to provincial considerations. Here, participants also talked about conditions in other places, to which they were connected by other people.

But this analysis also demonstrates that discourse about places is associated with a competitive, zero-sum approach to natural resources. When comparing places, speakers highlighted the potential winners and losers in a changing environment of less water in rural areas and more people in urban spaces. Rather than highlighting physical and social connections between human communities, place-based talk presented detached locations vying against one another. Talking about the environment from a position of discrete autonomy might strengthen a sense of place and social identification with one's community but can be counterproductive to the broader goals of environmental planning. Others have noted that environmental issues do not conform to human boundaries such as a county or state, and environmental decision making therefore must be organized around ecological units (watershed, aquifer) rather than social ones (Druschke, 2013). This analysis further complicates the dilemma, demonstrating that whether people are speaking of places defined by political boundaries or those with ecological demarcations, their talk conforms to neoliberal logics of property rights and market competition. Under such logics, nimby-like responses seem more likely.

In light of this dilemma, two best practices ought to be followed. First, conveners can creatively promote both provincial and more pluralistic considerations. In the case studied here, citizen-leaders decided to hold public meetings in many towns, reasoning that convenience would encourage more people to attend. But in emphasizing accessibility, conveners missed an opportunity to bring together people from different communities. In the only such instance in the transcripts, participants from different cities spoke of the relationship between their respective municipalities and their common water supply:

> SPEAKER ONE: I would personally appreciate it if El Dorado would take some of Wichita's water customers since you got plenty and we're kind of, you know . . .
>
> SPEAKER TWO: We would be glad to do that. [Group laughter]
>
> SPEAKER THREE: Seems like a no brainer to me, it's the same storage capacity and a lot more people are drinking out of Cheney. Water's great in El Dorado.
>
> SPEAKER ONE: If you look at Wichita from 20,000 feet, you have Cheney and El Dorado Lakes are equal distance . . . so it's very logical.

Speaker One makes his initial request apparently in jest, but as the discussants pursue the idea, they become more serious. This single example suggests that public discussion must include some diversity to encourage a "20,000 feet" view of communities and their shared stake in the environment.

To promote this perspective, officials might host local meetings along with regional gatherings—a complex task act made more feasible by digital technologies. In the process studied here, the Kansas Water Office made a Herculean effort to collect, transcribe, and post online all written comments from the meetings. But the asynchronous nature of this feedback likely limited its effectiveness. If local assemblies were instead held in distant places at that same time, participants could begin discussions face-to-face with their neighbors, then meet via communication technologies to hear other perspectives on the same topics.

Second, those convening environmental discussions must consider how their framings can influence the content and style of discussions. Here, state officials emphasized that a committee of residents designed the process to facilitate local decision making. They also stressed the importance of stakeholder input. A discourse that so strongly defers to indigenous autonomy encourages participation. However, this discourse fails to remind citizens that they are not alone and that public leaders must weigh their wants and needs against those of others. Ideally, participants in community-based processes would be approached symmetrically, indicating that their local knowledge is invaluable but incomplete (Lind, 2014; see also Peterson et al., 2006; Usher, 2013). Conveners therefore ought to balance deference to local preferences with talk of their own responsibility to promoting the common good.

Finally, this study speaks to a specific case. The sponsors of the process—the Kansas State Department of Agriculture and the Kansas Water Office—might have suggested to participants which topics were relevant. Additionally, the sponsorship likely brought more agriculturalists and their biases to these meetings. Therefore, future considerations should include the full gamut of ecological issues.

But at the same time, this research demonstrates the productive tension between localized knowledge and theoretical generalization in the scholarship of public participation. For if community-engaged scholars have a special obligation to "supralocal, institutional, and policy level-work that transcends communities," *as well as* a "moral imperative" to engage in "structural analysis and advocacy," then knowing cases intimately must be our charge (Hartman, Sanchez, Shakya, & Whitney, 2016, p. 166). Therefore, scholars ought to continue exploring how various communities, stakeholders, and leaders use language to tackle their shared concerns. In doing so, we likely will find useful special *and* general knowledge.

Placing Language in the Times

This study was motivated by a paradox of language and its use: people need to talk about places to foster inclusive decision making and marshal their own sense of being situated in a specific geography, but reference to place almost always excludes someone or something in some way. The study ends with more clarity as to what talk of place leaves out and why. It also proposes more investigation to overcome the limits of place, rather than abandoning its discourse. As Cahoone (2001, p. 3) cautions, "while neighborhood is not a sufficient condition for wider conversation . . . it is a necessary condition." And as Ron Arnett (2011, p. 632) observes, "modernity, with its commitment to universal truth, may wish to take the provincial or the local off the table of conversation," but this would be impossible "in an era defined by difference."

Speaking of modernity, this analysis also raises questions about the utility of place-based talk in organizing concerted community action. For while oral and print cultures maintain a "bond between physical place and social place," electronic media "lead to a nearly total dissociation of physical place and social place" (Meyrowitz, 1987, p. 115). Ironically, at the very moment when modern life has presented deeply complex environmental problems, modernity also threatens to alter one of our most important tools—sense of place—for addressing such problems. Apparently, the planet continues to need all that public servants, concerned citizens, and engaged scholars bring to bear.

Notes

1. Meetings were held in Salina, Wichita, Goodland, Hiawatha, Ottawa, Manhattan, Erie, and Beloit. Public participation varied from as few as 30 at the Wichita meeting to as many as 60 at the Goodland meeting.

2. Comments made by Kansas Secretary of Agriculture Jackie McClaskey at a March 2015, public meeting in Salina, Kansas.

3. As an affiliate faculty member in the Institute for Civic Discourse and Democracy, I was twice called upon to facilitate small-group discussions. Otherwise, I attended the meetings solely to observe and record the proceedings. I was introduced at each meeting as "a professor from Kansas State who studies public processes" and as the person responsible for the recordings. Transcripts from the groups I facilitated were excluded from this analysis.

References

Arnett, R. C. (2011). Civic rhetoric—meeting the communal interplay of the provincial and the cosmopolitan: Barack Obama's Notre Dame Speech, May 17, 2009. *Rhetoric & Public Affairs, 14,* 631–671.

Berry, W. (2003). *Citizenship papers.* Washington, DC: Shoemaker & Hoard.

Brocato, E., Baker, J., & Voorhees, C. (2015). Creating consumer attachment to retail service firms through sense of place. *Journal of the Academy of Marketing Science, 43*(2), 200–220.

Brooke, R. (2012). Voices of young citizens: Rural citizenship, schools, and public policy. In K. Donehower, C. Hogg, & E. E. Schell (Eds.), *Reclaiming the rural: Essays on literacy, rhetoric, and pedagogy* (pp. 161–172). Carbondale, IL: Southern Illinois University Press.

Burke, K. (1950). *A rhetoric of motives.* New York, NY: Prentice-Hall.

Cahoone, L. (2001). *Locale and progress.* Paper presented at the Annual Meeting of the American Political Science Association, San Francisco, CA.

Cantrill, J. G. (2012). Amplifiers on the commons: Using indicators to foster place-based sustainability initiatives. *Environmental Communication, 6,* 5–22. doi: 10.1080/17524032.2011.640703

Cantrill, J. G., & Senecah, S. L. (2001). Using the "sense of self-in-place" construct in the context of environmental policy-making and landscape planning. *Environmental Science & Policy, 4,* 185–203. doi:http://dx.doi.org/10.1016/S1462-9011(01)00023-5

Cantrill, J. G., Thompson, J. L., Garrett, E., & Rochester, G. (2007). Exploring a sense of self-in-place to explain the impulse for urban sprawl. *Environmental Communication, 1,* 123–145. doi:10.1080/00036810701642514

Carbaugh, D., & Cerulli, T. (2013). Cultural discourses of dwelling: Investigating environmental communication as a place-based practice. *Environmental Communication, 7,* 4–23.

Center for International Environmental Law for World Wildlife Fund-US. (1999, January). *Effective decision-making: A review of options for making decisions to conserve and manage Pacific fish stocks.* Retrieved from http://www.ciel.org/Publications/effectivedecisionmaking.pdf

Cocciolo, A., & Rabina, D. (2013). Does place affect user engagement and understanding?: Mobile learner perceptions on the streets of New York. *Journal of Documentation, 69,* 98–120. doi:10.1108/00220411311295342

Coleman, S., & Firmstone, J. (2014). Contested meanings of public engagement: Exploring discourse and practice within a British city council. *Media, Culture & Society, 36,* 826–844. doi:10.1177/0163443714536074

Daniels, S. E., & Walker, G. B. (2001). *Working through environmental conflict: The collaborative learning approach.* Westport, CT: Praeger.

Devine-Wright, P. (2009). Rethinking nimbyism: The role of place attachment and place identity in explaining place-protective action. *Journal of Community & Applied Social Psychology, 19,* 426–441. doi:10.1002/casp.1004

Druschke, C. G. (2013). Watershed as common-place: Communicating for conservation at the watershed scale. *Environmental Communication, 7,* 80–96.

Endres, D. (2012). Sacred land or national sacrifice zone: The role of values in the Yucca Mountain participation process. *Environmental Communication, 6,* 328–345.

Fisher, W. R. (1987). *Human communication as narration: Toward a philosophy of reason, value, and action.* Columbia, SC: University of South Carolina Press.

Habermas, J. (1984). *The theory of communicative action.* (T. McCarthy, Trans.). Boston, MA: Beacon Press.

Hart, R. P., Childers, J. P., & Lind, C. J. (2013). *Political tone: How leaders talk and why.* Chicago, IL: University of Chicago Press.

Hartman, E., Sanchez, G., Shakya, S., & Whitney, B. (2016). Critical commitments to community and campus change. In M. A. Post, E. Ward, N. V. Longo, & J. Saltmarsh (Eds.), *Publicly engaged scholars: Next generation engagement and the future of higher education* (pp. 156–168). Sterling, VA: Stylus.

Hess, K., & Waller, L. (2014). Geo-social journalism. *Journalism Practice, 8,* 121–136.

Ippolito, J. M. (2009). Words, images and avatars: Explorations of physical place and virtual space by Japanese electronic media artists. *Leonardo, 42,* 421–395.

Kemmis, D. (1990). *Community and the politics of place.* Norman, OK: University of Oklahoma Press.

Lane, R. E. (1962). *Political ideology: Why the American common man believes what he does.* New York, NY: The Free Press of Glencoe.

Lee, R. (1995). Electoral politics and visions of community: Jimmy Carter, virtue, and the small town myth. *Western Journal of Communication, 59,* 39–60.

Lind, C. J. (2014). Deference in the district: An analysis of congressional town hall meetings from the C-SPAN video library. In R. Browning (Ed.), *The C-SPAN archives: An interdisciplinary resource for discovery, learning, and engagement* (pp. 59–77). West Lafayette, IN: Purdue University Press.

Marzorati, R. (2013). Imagined communities and othering processes: The discursive strategies of established Italian residents in a Milan city neighbourhood. *Journal of Language & Politics, 12,* 251–271. doi:10.1075/jlp.12.2.05mar

Meyrowitz, J. (1987). *No sense of place.* New York, NY: Oxford University Press.

Milligan, M. J. (1998). Interactional past and potential: The social construction of place attachment. *Symbolic Interaction, 21,* 1–33. doi:10.1525/si.1998.21.1.1

Nabatchi, T., & Amsler, L. B. (2014). Direct public engagement in local government. *The American Review of Public Administration, 44*(4 Supplement), 63S–88S. doi:10.1177/0275074013519702

Peterson, T. R., Peterson, M. N., Peterson, M. J., Allison, S. A., & Gore, D. (2006). To play the fool: Can environmental conservation and democracy survive social capital? *Communication & Critical/Cultural Studies, 3,* 116–140.

Polletta, F., & Lee, J. (2006). Is telling stories good for democracy? Rhetoric in public deliberation after 9/11. *American Sociological Review, 71,* 699–721. doi:10.1177/000312240607100501

Procter, D. E. (1996). Placing Lincoln and Mitchell counties: A cultural study. *Communication Studies, 46,* 222–233.

Rice, J. (2012). From architectonic to tectonics: Introducing regional rhetorics. *Rhetoric Society Quarterly, 42,* 201–213. doi:10.2307/41722431

Sanderson, M. R., & Frey, R. S. (2014). From desert to breadbasket . . . to desert again? A metabolic rift in the High Plains Aquifer. *Journal of Political Ecology, 21,* 516–532.

Seggern, M. V., Merrill, A., & Zhu, L. (2010). "Sense of place" in digital collections. *OCLC Systems & Services: International Digital Library Perspectives, 26,* 273–282. doi:10.1108/10650751011087639

Shellabarger, R., Peterson, M. N., Sills, E., & Cubbage, F. (2012). The influence of place meanings on conservation and human rights in the Arizona Sonora borderlands. *Environmental Communication, 6,* 383–402. doi:10.1080/1752 4032.2012.688059

Thomson, C., Bennett, D., Johnston, M., & Mason, B. (2015). Why the where matters. *Pacific Journalism Review, 21,* 141–161.

Tracy, K. (2001). *Rhetorically-informed discourse analysis: Methodological reflections.* Paper presented at the Alta Conference on Argumentation. Retrieved from http://search.ebscohost.com/login.aspx?direct=true&db=ufh&AN=20760432&site=ehost-live

Tversky, A., & Kahneman, D. (1981). The framing of decisions and the psychology of choice. *Science, 211,* 453–458. doi:10.1126/science.7455683

Usher, M. (2013). Defending and transcending local identity through environmental discourse. *Environmental Politics, 22,* 811–831. doi:10.1080/09644016.2013.765685

Walsh, K. C. (2012). Putting inequality in its place: Rural consciousness and the power of perspective. *American Political Science Review, 106,* 517–532. doi:10.1017/S0003055412000305

Ward, S., & van Vuuren, K. (2013). Belonging to the Rainbow Region: Place, local media, and the construction of civil and moral identities strategic to climate change adaptability. *Environmental Communication, 7*(1), 63–79. doi:10.108 0/17524032.2012.753098

Zarlengo, T. (2016, June 4). *Frankenbug meets the Conch Republic: Engagement, expertise, and "strategic irrationality" in public scientific controversies.* Paper presented at the 5th Iowa State University Summer Symposium on Science Communication, Ames, IA.

Chapter 5

Cultural Discourses of Public Participation

Insights for Democratic Design and Energy System Transformation

Lydia Reinig and Leah Sprain

Shifting the nature of how energy is used, delivered, produced, or sourced within a city—energy transition (Araújo, 2014)—could easily be seen as a matter of electrical wires and grids, wind turbines and hydroelectric generators. Indeed, infrastructure networks play a fundamental role within sustainability transitions (Bolton & Foxon, 2015). Nonetheless, energy infrastructure is deeply embedded in broader sociotechnical contexts (Goldthau, 2014). Transformation of technical systems is not determined by scientific, technological, or even economic rationality; energy system transformation is fundamentally social, political, and cultural (Bolton & Foxon, 2015). Yet the participatory processes that support energy transition at the community level are still being developed. Energy communication research has focused on discourses of decision making about energy in the context of crises, considering how various stakeholders impact decisions about energy technologies (Endres, Cozen, Barnett, O'Byrne, & Peterson, 2016). This has taken the form of examining public positions about energy issues (e.g., stances on nuclear power) or public understanding of new energy technologies (e.g., smart grids). Rather than focus on public opinion about specific energy options, we consider the public's role within energy transition. How should the public participate in energy system transformation? What forms of public engagement does energy governance require? What public

121

participation infrastructure is needed to support a just energy transition? What does energy democracy entail?

We introduce big questions about the principles and practices of public participation within energy system transformation despite being acutely aware that these questions will not be fully answered in this chapter. Nonetheless, these questions guide our inquiry, establishing the need to better understand the possibilities for and requirements of public participation within energy system transition. Given the dissatisfaction, distrust, and incivility that characterize conventional public participation (Natabtchi & Leighninger, 2015), meeting these needs by simply replicating public participation practices from other environmental arenas is insufficient. Instead of immediately turning to innovative practices (e.g., deliberation, collaborative learning, joint fact finding), we examine cultural discourses about public engagement active within a community as a foundation for designing public participation and theorizing energy governance. This move reflects our argument that examining cultural discourses can generate insights about how to design more meaningful public engagement that shapes the future of a community's energy system.

A cultural approach to analysis develops knowledge from the perspective of the community; however, culture is not synonymous with community. Drawing from the ethnography of communication (Hymes, 1974), culture is defined as "a system of meanings, an organized complex of symbols, definitions, premises, and rules" (Philipsen, 1992, p. 14). The notion of cultural discourses is derived from Cultural Discourse Analysis (CuDA), which argues that locally situated meanings and taken-for-granted premises for interpretation are embedded within a community's everyday talk (Carbaugh, 2007). Adopting a cultural approach allows for tracing locally available means of public participation and their meanings from everyday talk. Isolating the range of means and meanings allows us to analyze how cultural discourses configure participatory practices.

Cultural discourses provide insight into public participation in three overlapping ways. First, public participation varies across cultures, influenced by the organization of government structures and notions of who can participate and how (see Sprain, 2006 for review). What citizens count as public participation also varies; for example, participation in the Young Communist Union in the Soviet Union was not considered political participation, while Chinese "work-go-slows" proved a novel form of public participation. Cultural discourse analysis can help reveal the locally active forms and meanings of public participation that should not be presumed to be universal across communities.

Second, cultural variety can result in significant gaps between how different groups engage in public participation. Sprain, van Over, and Morgan (2016) traced cultural meanings for participation in two environmental decision-making processes, arguing that multiple, and at times contested, meanings and premises emerge within the same public participation process. Tribal communities maintained cultural premises "that one cannot speak and act freely in a documented public space with unknown others" (p. 256), which clashed with expectations that community members would speak on camera at broadcasted public meetings. Cultural discourse analysis can help prevent introducing public participation practices likely to fail given inherent clashes with the cultural system. For example, Dean (2016) cautions that introducing agonistic procedures into solidaristic institutional cultures may result in alienation.

Attentiveness to local means and meanings of participation can also be leveraged toward intervention—our third reason for conducting cultural discourse analysis. Through cultural discourse analysis, scholars and practitioners develop a richer understanding of local communication practices, people's social worlds, and the environmental actions that best reflect them (Morgan, 2003). Thus, cultural discourse analysis may provide inventional resources (Schwarze, 2006) for developing relevant, appropriate, and meaningful public participation processes built from cultural practices. Cultural knowledge is not simply a matter of minimizing cultural clashes but instead offers the basis for drawing on cultural knowledge to design public participation from locally relevant forms of strategic action (Sprain & Boromisza-Habashi, 2013).

Cultural knowledge is a valuable input because designing meaningful public engagement is a deceptively difficult task. Environmental governance scholars widely recognize a significant gap between ideals of public engagement in environmental decision making and actual processes and practices for participation (Bulkeley & Mol, 2003). While many studies of public engagement seek to name and assess the gaps (e.g., Reed, 2008; Newig & Fritsch, 2009), fewer studies thoughtfully engage with ways situated talk reveals local premises for public participation that, in turn, suggest possible ways to make public processes and practices more meaningful to and effective for participants. Cultural discourse analysis positions scholars and practitioners to contribute to designing meaningful public participation and theorizing energy democracy.

To support this argument, we examine cultural discourses of public participation within a community undergoing self-described energy system transition. We briefly review the literature on public participation in environmental governance before introducing our case study—Boulder's energy future. Using CuDA as a framework, analysis is organized around five forms of public participation active within our case study and their associated meanings. From this analysis, we suggest how cultural discourses can inform the design of public participation before identifying broader implications for energy system transformation and energy democracy.

Public Participation in Environmental Governance

Public participation is a broad concept that gets referenced through multiple, overlapping terms, such as community engagement, stakeholder participation, citizen involvement, public deliberation, collaborative problem solving, and political participation. Given this variety, public participation is "an umbrella term that describes the activities by which people's concerns, needs, interests, and values are incorporated into decisions and actions on public matters and issues" (Nabatchi & Leighninger, 2015, p. 14). In turn, public participation has multiple meanings that can reflect differing normative conceptions of social organization and citizens within it (Dean, 2016). For example, the empowered self-interest of the consumer-citizen and the other-oriented citizen of deliberative democracy can both be understood as forms of public participation. Scholars have developed typologies for forms of public participation that can be placed along a continuum from least to most legitimate or categorize public participation by political ideology (see Dean, 2016 for review). This impulse to build typologies reflects a desire to simultaneously recognize a range of forms of public participation while noting that they reflect significant differences in design features that influence the meaning, legitimacy, and outcomes of public participation (Fung, 2006).

Nabatchi and Leighninger (2015) provide an alternative typology with three overarching categories of participation: thick, thin, and conventional participation. Thick participation enables people to work together in small groups to learn, decide, and act through processes such as deliberation, action planning, and design charettes; thick participation is generally considered the most meaningful and powerful form of participation. Thin participation activates people as individuals to participate by sharing their ideas in ways that take only a few minutes, such as signing petitions, liking a cause on Facebook, or ranking ideas in a crowdsourcing campaign. Conventional participation includes established institutional forms like public hearings that were developed to provide citizens with checks on government power.

Environmental communication scholars have long described, inter-preted, and critiqued conventional public participation practices because these processes are mandated any time a project with environmental impacts is proposed. Conventional participation often frustrates citizens and public officials alike, increasing feelings of citizen powerlessness, discouraging offi-cials who have to deal with hostile citizens, and contributing to the belief that public participation actually degrades the quality of decision making (Nabatchi & Leighninger, 2015). Public hearings, for example, are often criticized for creating distance between parties and limiting diverse stake-holder input, contributing to conflict-ridden interactions and adversarial positioning (Walker, 2004; Wills Toker, 2004; Senecah, 2004; 2007; Buttny & Cohen, 2015). Moreover, scholars recognize that simply having oppor-tunities for participation does not mean that participation is consequential for environmental decision-making processes (Depoe & Delicath, 2004).

MAKING PARTICIPATION MORE MEANINGFUL TO THE PUBLIC

Environmental communication scholarship has theorized the challenges and failures of public participation as dynamic communicative processes (see Depoe & Delicath, 2004; Senecah, 2007), in turn reimagining, configur-ing, and evaluating innovative public participation structures (e.g., Walker, 2007). This scholarship asks: How can we come to better envision and assess "meaningful" processes for citizen participation in environmental decision making? While this question might suggest the development of measures of effectiveness, notions of "meaningful" participation are contextual—dependent on purposes, expectations, and outcomes of participation (Depoe & Delicath, 2004). As meaningfulness is situated, multiple, divergent meanings for public participation can be simultaneously activated in talk (Lassen et al., 2011). Government agencies might assess the process as meaningful based on the number of meetings held or public comments received, yet the public often finds these measures insufficient (i.e., far from "meaningful") (Walker, 2004). These various ways of understanding participation help remind scholars that participation does not exist in a natural state—participation is an emergent and coproduced phenomenon through the performance of collective partic-ipatory practices (Chilvers & Longhurst, 2016).

Environmental communication scholarship tends to turn its attention toward reimagining processes by developing theoretical frameworks and concepts that foster critical reflection (e.g., Senecah, 2004; Walker, 2004; Martin, 2007). Rather than develop normative expectations for judging good and bad participation, orienting to cultural discourses provides important insights into the *process* of designing public participation through an empirically oriented exploration of how diverse forms of participation get made. Many public participation practitioners do research before initiating a public participation process. Deliberative inquiry (Carcasson & Sprain, 2016), for example, starts with deliberative issue analysis, which can include analyzing public discourse, stakeholder interviews, open-ended surveys, and focus groups. In addition to doing situated research about the issue, practitioners would benefit from considering local forms and meanings of public participation. This does not, of course, preclude developing new forms of public participation; intervention and public participation innovation can be the goal from the outset. Nonetheless, understanding the existing social system can provide useful guidance for public participation practitioners.

Boulder's Energy Future

The city of Boulder, Colorado, has been taking steps for more than a decade to address climate change. In 2002, Boulder voters passed a resolution to uphold Kyoto Protocol goals and reduce emissions by seven percent from 1990 levels by 2012. The city took many innovative actions—developing extensive recycling and composting programs, expanding the bicycle trail system, and instating the nation's first carbon tax (City of Boulder, 2015)—yet could not sufficiently reduce carbon emissions without transforming the city's energy system.

Boulder's private energy provider, Xcel Energy, maintains a carbon-intensive energy portfolio. Starting in 2005, the city began studying other options for power generation; in 2010, the city let the franchise agreement with Xcel expire while continuing negotiations about whether Xcel could reach the city's energy goals. Simultaneously, the city began exploring the creation of a municipal power utility. In 2011, Boulder citizens voted on two measures that authorized municipalization efforts to continue. Measure 2B increased the utility occupation tax, allowing the city to raise nearly 2 million dollars by 2017 toward starting a power utility (*Boulder Weekly* Staff, 2011). Measure 2C amended the city charter to allow for continued exploration and creation—if deemed feasible—of a municipal utility. Boulder has subsequently explored developing a muni through expert studies that surveyed environmental, financial, legal, and technical dimensions. A ballot measure that would have jeopardized municipalization was defeated in November 2013, and in the following May, the City Council created a local power utility in the charter. For the next three years, limited civic action occurred as the project moved through various legal and regulatory processes. In April 2017, Xcel put forth two settlement proposals—a partnership agreement between Xcel and the city or an immediate buyout of Xcel's assets—which Boulder's city council voted not to accept (Burness, 2017a). After long-awaited deliberation, on August 30, 2017, the Public Utilities Commission issues a series of decisions that provided Boulder with a "path forward" to pursue a municipal utility. Later that year, Boulder voters narrowly passed an extension of the utility occupation tax, which funds continued exploration (Burness, 2017b).

Given our contention that cultural discourses can inform the design of public participation, we note that the city has explicitly expressed interest in expanding the range of innovative public engagement approaches it uses. This case study holds practical significance for the authors as a preliminary step toward working with the city and its publics to design more innovative forms of public engagement that also connect with communities and publics not currently reached by the city. We also have experience as deliberative practitioners designing public engagement on a variety of environmental issues (see Carcasson & Sprain, 2016; Sprain, Carcasson, & Merolla, 2014; Sprain & Carcasson, 2013). Whereas this chapter is written before implementation of our public processes, our ongoing practitioner work helps situate us within Boulder and this research.

Methods

Our analysis focuses on public comments and council discussions during city council meetings on Boulder's Energy Future from 2010 to 2016. Our corpus includes 32 transcribed meetings totaling more than 800 single-spaced pages. We have also conducted fieldwork at city council meetings, public meetings on energy issues, and community and campus events related to energy that informs our interpretations. We have read the local paper and looked at thin forms of public participation, such as Inspire Boulder, an online public participation incubator (see Sprain, 2014). Our focus on council meetings as a site of public discourse reflects our sense that these meetings best embody formal decision making about energy to date. If forms of public participation are not discussed during these meetings, we presume that they have less significance and influence on decisions. For example, Inspire Boulder is mentioned only *once* in more than 800 single-spaced pages of transcripts as part of a report about how a city staff member is "trying to reach out to different parts of the community," making us reticent to formulate cultural claims about it. It is, of course, likely that there are forms of public participation within Boulder that do not appear in our meeting data. Our data are surprisingly scant on criticisms of public participation, even though we know that there are some significant concerns within Boulder about the City Council. This is an important limitation of focusing analysis on council meetings.

ANALYTICAL FRAMEWORK

In CuDA, an analyst describes cultural forms and their situated meanings by focusing on the range of interpretations for a focal practice that are active within a community's talk, focusing specifically on meanings related to identity, actions, relationships, affective orientations, and dwelling (Carbaugh, 2007; Carbaugh & Cerulli, 2013; Scollo, 2011). These interpretative accounts draw on a rich set of concepts from the ethnography of communication, including attention to key terms, cultural propositions (e.g., definitions, premises, beliefs, or values), cultural premises, semantic dimensions, and norms that are embedded in everyday communication.

We set out to ask: What cultural discourses of public participation are active within Boulder's energy municipalization process? What forms of public participation provide key means of communicative action in Boulder, and what are their meanings? Through these questions, we seek to better understand the connections between communicative action and participatory ideals that can, in turn, aid in the design of strategic action.

Analysis

Our analysis of city council meetings suggests five prominent forms of public participation: attending and speaking at council meetings, email between citizens and council members, city-initiated education and outreach efforts, working group membership, and voting. Abbreviated analysis of the first three forms serves to situate the range of locally relevant forms of participation, whereas an extended discussion of working groups and voting highlights a range of meanings that configure each participatory practice.

ATTENDING AND SPEAKING AT CITY COUNCIL MEETINGS

Given the widely institutionalized practice of public comment periods during public meetings, it comes as no surprise that attending and speaking at city council meetings is a prominent means of participation. Whereas public comment sessions reify relatively rigid dynamics for interaction that limit participation (as is well established in the literature), our analysis underscores how attending and speaking has cultural meaning for participants. At the beginning of every public comment period, the mayor provides "the rules for decorum," which outlines policies that limit extended public comments:

> [L]et's go to the public hearing, um thanks for being so patient, . . .
> So there are at the moment 50, as in 5-0, people signed up,
> um everybody will get their—up to two minutes, if you can
> be a little shorter that's great, we will definitely hear from all
> of you, . . . everybody gets two minutes, and lemme just say
> that um I'm really gonna cut you off at two minutes, um and
> we're gonna be very strict about this, and that's simply a mat-
> ter of fairness, I mean the easiest way to be absolutely fair to
> everybody is to treat everybody the same rotten way [((laughs))]
> which is to—which is to cut you off when the buzzer sounds.
> (Council 4/16/13)

The intent is to be "fair to everybody" and "treat everybody the same rotten way," suggesting some acknowledgement of public hearing criticisms while also reinforcing the time boundaries for participation. The mayor continues to articulate the rules for participation:

> Other than that, again please follow the rules of decorum, respect
> what people are saying, we don't need applause, we don't need
> boos, we get it, we understand who's in the audience, that's fine,
> we're delighted to have a large group of people from diverse
> communities and with diverse interests, it's exactly what we
> wanted. (Council 4/16/13)

By reiterating "the rules of decorum," the mayor constrains the ways speakers can interact with the council and with each other, minimizing the theatrical, performative nature of addressing the council in front of a public audience. In turn, speakers are constituted as having diverse, conflicting perspectives that need to be managed—the sort of diversity that is desirable within public participation.

Despite the explicit framing that the council is interested in hearing from the public, statements made during public comment sessions prompt limited, if any, interactions between the council and the public; immediate uptake (e.g., acknowledges, responds to, and/or asks follow-up questions) is rare, occurring only when a new perspective is offered or technical information requires clarification. These patterns broadly fit the observations in the literature on the constraints of public participation within public meetings (see Buttny & Cohen, 2015).

EMAIL

When explaining public comment rules, the mayor says, "I mean we know we're cutting you short with two minutes, if you have longer comments, please email them, council absolutely reads email" (Council, 4/16/13). While in practice such a statement functions to preclude extended individual comments during meetings, suggesting that citizens email the council frames speaker-listener relationships in which the mayor assures the public that the council is available and responsive. Council members make comments—such as, "I've gotten a lot of emails, and I think you all have gotten probably the same ones I've gotten uh with the concerns from some of the uh industrial users" (Council, 11/5/12)—that reference interactions with relevant stakeholders and suggest their concerns are being heard (or represented) while also allowing council members to reinforce their own positions but frame them as public concerns. Email is lauded by the City Council as a means for the public to contribute in ways that enable listening but simultaneously can close conversation by reinforcing council members' discretion to respond.

OUTREACH AND EDUCATION

The city seeks to include the public in Boulder's energy future discussions through "education and outreach." Two primary means for outreach are roundtable discussions and public presentations, which offer the public access to technical explanations, and opportunities to discuss issues related to municipalization and climate change (e.g., "it'll be a roundtable where you—you will get a presentation and there'll be an opportunity for public to respond" [Council, 11/15/12]).

Talk about outreach and education underscores the expectation that the public should be educated about and included in energy system transition discussions and decisions. The city has the responsibility to share requisite information in accessible ways (e.g., "citizens are given full access" [Council, 6/7/11]) so that the public can consider "risks and benefits" and make "informed decisions" about municipalizing. To a degree, education and outreach efforts presume that the public initially lacks knowledge, yet educating the public also presumes that "they care" about transition and are prepared to "listen, understand, and decide what's acceptable to them." Boulder citizen Julie affirmed citizens' involvement:

> People are becoming aware of where their energy is coming from, and they care . . . At this point, we don't have enough information to make an informed, fact-based emotional-non-emotional decision. I hope for clean, green, locally sourced renewable energy for Boulder. I have high hopes because this is Boulder, where citizens want to know and want to be involved. Please make sure that these hopes are considered, and that the related details are spelled out in the Xcel proposal so that the citizens of Boulder can listen, understand, and decide what's acceptable to them. (Council, 6/7/11)

Tailoring outreach efforts toward these interested, invested citizens and offering educational opportunities that build on existing public understanding and commitments are then essential.

Municipalization is framed as an effort that must be done in line with citizens' values and support. This is illustrated in council member Macon's statements:

> . . . I mean when we have talked about a municipalization, we've—we have recognized the importance of outreach to the community so that we move in—in step with the community, changing the—the habits that people have about using electricity and how they um and—and what sources they get it from. (Council, 6/7/11)

By positioning the need to "move . . . in step with the community," council member Macon recognizes that energy system transition requires coordinated action between individuals and governments. While the government can seek to change the utility infrastructure (e.g., supply) and pass ordinances (e.g., rate structures), making progress toward climate action goals requires the public to both support government initiatives and change their own consumption habits. Macon also recognizes transformation as a slow, time-bound process; it cannot be rushed without risking alienating the public's support, which makes outreach important.

WORKING GROUPS

To date, thirteen working groups (also called task forces) have been created by the city to advise the council on topics such as rates, governance, resource modeling, solar, energy services, reliability and safety, and Boulder-Xcel partnership. As the working group states on its website, "The Energy Future Project has formed volunteer community working groups to obtain input from residents, businesses, and stakeholders on project focus areas" (City of Boulder, 2016). The process for appointing working group members varies by topic, but, generally, participants should be able to "represent Boulder community interests, the residents' interests," and have relevant topic expertise—the previous quote continues "as well as have a good perspective on what the constraints of the laws are" (Council 2/4/14). Working groups meet (often with a facilitator and city staff) over several months to discuss issues identified by the council and to generate a public report that is presented to the council. Solar Working Group member Yael captures its overall purpose:

> [W]e formed this working group to both, um, address concerns that were out there in the industry, locally, and well as look at what the future of solar could be in Boulder with or without a municipal utility. Um, this group did a tremendous amount of work in a very short time . . . They developed a huge matrix of opportunities, um, that came about from a few brainstorming sessions, and then really distilled that down to form the recommendations that are in front of you tonight. (Executive session, 8/19/14)

Council, staff, and working group members' talk characterizes their interactions as what Nabatchi and Leighninger (2015) would call thick participation. Working groups brainstorm, prioritize, and, ultimately, make recommendations for the council's consideration. These sessions should be "open and honest," providing opportunities to discuss multiple perspectives. Yet the council disagrees about some of the procedures for the working groups, including whether consensus is necessary. Council member Suzanne's statements—made during a discussion about the working group's function—provides several premises about the ideal form and purpose of the working group's report to the council:

> I think what wouldn't be helpful from this group is to have a vote on particular options, what would be useful from this working group is their expertise. There are a bunch of smart people that represent a bunch of different perspectives, they're not all the perspectives, but it could be useful information, and so I guess I just wanted to clarify that I like the no votes, no quotes thing, I want their expertise, I want their analysis on whatever option or couple options that they think are really worth talking about, um and that's what would be useful to this group who ultimately is the one that makes the votes, and not theater, not drama about votes and that sort of thing, not headlines that are a distraction, but their best thinking. (Council, 4/16/13)

Working groups should not be about "theater" or "drama" or "votes" or "headlines that are a distraction"; instead, working groups should be about "their best thinking" and "their analysis on whatever option or couple of options that they think are really worth talking about." By considering options, working groups can provide useful input to decision making, yet working groups do not have any decision-making authority.

The City Council maintains authority over both making decisions and directing working groups. When a particularly politicized working group on the Boulder-Xcel partnership decides that it doesn't need a moderator to facilitate the discussions, council member Lisa reasserts the council's control over the process:

> Uh, one of the things that you said that is of concern to me is that the task force itself decided that they didn't think a moderator was necessary . . . I feel like this is a city, . . . a city run process. And I think there's some things that the city and the city council needs to make the decisions, and not the task force. It in some ways, it seems like the task force is a runaway train. And, and that has me very concerned, and so, I, I want these decisions to come back to the council. I, I think we do need to have a moderator. (Council, 2/4/14)

This is just one instance of where the council maintained its control over the working group process. As council member Lisa stated, "I don't want to see [working groups] in the driver's seat. The city and the city council should be in the driver's seat" (Council, 2/4/14).

The council maintains control and decision-making power while looking to working groups to provide relevant expertise on issues. Council and working group members alike celebrate the depth of expertise within the community that, in turn, supports the working groups. Solar Working Group member Yael continued addressing the council by introducing another working group member:

> He has 17 years of professional experience, um, 10 in broad, more broadly in energy and 7 more specifically in solar. He's held roles in project, um, finance and project development for solar companies since 2007. Um, his career highlight was the development of a 30-mega watt solar project that was commissioned in two thousand and eleven. He has an MBA from Georgetown and a degree from University of Colorado in mechanical engineering, so here's John. (Executive session, 8/19/14)

John is introduced by multiple different credentials—years of experience, project roles, project outcomes, and his academic degrees. Providing all of this information is not heard as inappropriate or excessive; working groups are a form of public participation where expertise is deeply valued. As Mayor Matt remarks after the presentation:

> The report was quite remarkable . . . I think this community needs to recognize the amazing amount of talent that we have out there, the expertise we have. Um, Boulder, Boulder is, uh, kinduva quite amazing place when it comes to stuff like this. We just got folks who are nationally and internationally known experts of all of these areas. And they are incredibly generous with their time and their skills and their knowledge, um, offering them up to the city. As Sam said, you know, you could pay a lot for a professional report and it wouldn't necessarily be any better than what we get, um, with our staff and working groups. (Executive Session, 8/19/14)

The favorable comparison to professional consultants is telling: working groups are valued because of the technical advice they provide. A broad range of experiences and perspectives is important, but participants are not seen primarily as citizens or members of the public: working group members are experts. As a form of public participation, working groups should be composed of experts who can work through complex issues, discuss diverse perspectives, and make recommendations to the council while the council maintains discretion and control over any decisions and the working group process.

VOTING

Discourses about voting suggest that voting serves multiple communicative functions (i.e., confirmatory vs. advisory) and possesses incongruent meanings (i.e., clearly stated vs. open to interpretation). Most commonly, voting is considered necessary for decision making because it allows the public to mandate a course of action and holds the council accountable. In this formation, voting acts as a conventional form of participation (Nabatchi & Leighninger, 2015). Voting serves a confirmatory function; it is a necessary democratic process that allows citizens to express support or disapproval of government actions, as suggested in this public comment from Boulder citizen Buzz:

> I think obviously this is going on the ballot, it should go on the ballot. I think the city's done as much work as it can, and now it's time to bring it to the people and see what they think, and I just hope that the ballot language reflects the choice very clearly and very honestly, because it is the biggest issue we ever are gonna face, probably in our lifetimes. So my only request to you folks on council and staff is to make sure the language really represents the magnitude of this decision, and the honesty—honestly reflects what is at stake so that we know what we're getting into and people who read it and haven't studied as much as the rest of us have can instantly understand it and make an—as informed decision as possible. (Council, 8/2/11)

Statements such as "it should go on the ballot" and "it's time to bring it to the people and see what they think" invoke voting as a means to present the issue to the general public so they can make "the choice." Concern remains that citizens are prepared for this important role—that they recognize the "magnitude of this decision" and are able to make an "informed decision" even if they "haven't studied as much as the rest of us." By emphasizing "that the ballot language reflects the choice very clearly and very honestly," voting is characterized as providing specific parameters to mandate action.

References to citizens voting do not end with the passage of initiatives 2B and 2C in the November 2011 election. The 2011 vote is leveraged as a rhetorical resource to argue that the public supports ongoing municipalization efforts and that the City Council and staff are accountable for carrying out mandated actions. Boulder citizen Ruth addressed the council:

> [I]n 2011 the voters approved ballot issues to 2B and 2C for exploring the possibility of creating a municipal utility, and along with that was approval for certain performance targets, listing the conditions necessary for moving forward. This language is now part of the city charter. . . . [I] Remind you that the voters approved a specific list of performance targets in 2011, and if opponents want to further extend the performance targets, they will need to do so in the charter or convince the majority of council to vote against possible munity formation on the basis of criteria not in the charter. . . . And remember, the vote might have been close in 2011, but the 2B 2C campaigns won, and we have an obligation to the voters to carry through with their vote to explore municipalization. (Council, 11/15/12)

Ruth's comments leveraged the 2011 public vote as confirmation of the public's support. The voters approved "performance targets" that are "now part of the city charter" serving as specific criteria and benchmarks; they are explicit and confirmed by the public to authorize continued efforts toward municipalizing.

A second meaning of voting is also active in these meetings: voting provides advice to governing bodies during decision-making processes. Voting has an advisory function when members of the public argue that additional votes are necessary to gage ongoing public support, clarify the meanings of charter language adopted in 2011, and run more inclusive processes that would allow Boulder County residents—who would be part of the utility service area but were excluded from the 2011 vote of Boulder citizens—to participate. Participants' talk suggests that voting can also function as thin participation—providing low-cost opportunities for a critical mass to express their positions (Nabatchi & Leighninger, 2015).

In one episode where voting is constructed as advice rather than confirmation, City Council was voting on whether or not to authorize city staff to continue exploring municipal utility formation. The public comment period lasted more than two and a half hours, with 44 people speaking both for and against continued exploration. Whereas speakers who supported moving forward leveraged the 2011 vote as a measure of public support (confirmation), skeptical participants called on the council to engage with the public further before taking additional actions, suggesting an advisory vote by citizens would provide an updated indicator of support. Angelique, a business community representative, provided a detailed proposal to put an advisory vote in front of Boulder citizens before the City Council could act further: "In terms of the vote I just wanted to clarify we didn't necessarily mean to suggest it would be a vote which would be contingent, um in other words perhaps it would be an advisory vote" (Council, 4/16/13). While an advisory vote offers the public an additional opportunity for participation, functionally, the influence of this participation could be limited because the government would not be accountable to it.

No matter its communicative meanings, voting is seen as a valuable form of participation—particularly for Boulder County residents and the unincorporated area of Gunbarrel. These groups would be included in Boulder's municipal utility service area but could not vote because they live outside city limits (i.e., "the 5800"). Regardless of whether they support or oppose municipalization, many of those disenfranchised come before the council to question on what grounds Boulder presumes to move forward without allowing non-residents to vote. Diana, a Gunbarrel resident, addresses the council:

I am in favor of renewable energy, [but] I am very disturbed by uh the fact that I never got a chance to vote. I think at a minimum a survey should be taken of the voters of the area, or the property owners, or the people who pay utilities. . . . Um you know if you're gonna have a vote, then schedule it in the 2014 or the 2016 vote, and really get a serious commitment from the other people. I don't think condemning the Xcel um utility at this time as taking into account um true impact on all the surrounding counties, I'm within the 5800, but you're gonna impact those outside. You have not consulted them, you don't know what impact it's gonna take on their rates . . . I really am in favor, I mean my husband's an environmental scientist, we are killing the climate. And I'm in favor of what you're doing, [but] I'd like a chance to vote on it. I'd like a chance for you to hear the people that are voting on it. (Council, 4/16/13)

The "chance to vote" is constructed as a matter of having sufficient opportunities for participation and actively engaging affected users as represented voters. Doing so ensures that the City Council is acting in ways that are accountable to the entire customer base—not just Boulder citizens. Beyond challenging the democratic ethos of Boulder, Diana's comments underscore that if the city seeks to actively support public participation, it cannot exclude people. For Gunbarrel and Boulder County residents, an advisory vote offers the potential for inclusion (or at least creates a sense of being considered).

Discourses of voting further imply expectations for a democratically governed electric utility. As Kate, a Boulder resident, says:

> But when Xcel signed us up for 16 more years of coal, they did not ask for my vote. However, here in Boulder, we did vote. I urge you to waste no time in moving toward a democratically accountable municipal electric utility. (Council, 4/16/13)

This statement highlights that a "vote" authorizing continued exploration of municipalization is also considered public participation in ongoing efforts that allow the public to make decisions about the future of their energy systems. Like Kate's call "to waste no time in moving toward a democratically accountable municipal electric utility," Ashwin, another Boulder citizen, calls for a democratically governed municipal utility, where the public has influence:

> Xcel and the publically [sic] sanctioned monopoly model undermines democracy. We do not elect the Public Utilities Commission that regulates Xcel, and they don't allow public comment like this, like we can do with our city. Um Xcel won't just resist decommissioning coal, because they haven't fully uh profited off of their assets, but they will actually stifle the democratic process that allows us to select our energy sources in democratic fashion. (Council, 4/16/13)

Speaking about a democratically governed municipal utility as superior provides a foil to the corporate model and the governing commission, both of which make decisions without consultation or accountability to the public.

Discussion

This volume expands the scope of community engagement theorizing through case studies that innovate participatory practice. Our chapter is poised at the intersections of considering best practices for specific participation designs and developing insights for expanding community engagement unique to energy democracy. We posit a cultural orientation to mapping existing cultural forms of public participation. Mapping the range of existing forms gives us a way to determine which practices are ripe for innovation. Rather than turning to outside theory for designing engagement and assessment, expanding community engagement boundaries starts by understanding the local significance of existing practices on their own terms. A cultural approach then enables design interventions that are accountable to local understandings and norms.

This chapter seeks to trace cultural discourses of public participation in energy system transformation active in Boulder by examining city council meetings. These forms of public participation active in Boulder should not be presumed to exist in other American cities, nor do their meanings necessarily transfer. Nonetheless, our findings should resonate (Tracy, 2010) to other communities considering energy system transition as well as provide a method for considering the cultural dimensions of public participation that cuts across environmental policy and climate action issues. Attentiveness to shared means and meanings—the crux of a cultural discourse approach—elucidates locally available forms of participation. Moreover, attentiveness to shared means and meanings situates norms for interpretation of participatory practices that can configure future public engagement processes and inform theorizing energy democracy.

Cultural Insights for Designing Public Participation

Our analysis of city council meeting discourses is structured around five forms of public participation for energy system transformation: attending and speaking at city council meetings, email contact, education and outreach, working groups, and voting. Most of these forms fit what Nabatchi and Leighninger (2015) call conventional participation—participation developed to provide citizens checks on government power. Only working groups provide a form of thick participation where groups can learn together, decide, and act, yet this form of thick participation is undermined by its current focus on narrow forms of technical expertise. When voting is associated with government accountability, it functions as conventional participation, yet voting is also presented as thin participation when it takes on an advisory role. The Boulder case calls for a broader range of public participation practices, including more thin participation opportunities to increase the quantity of public participation by a broader range of stakeholders and thick participation to increase the quality of public participation that can develop shared ownership of public action as well as reworking conventional practices like voting in response to public challenges about who has been excluded.

Cultural discourse analysis can inform design of public participation to address these needs in three overlapping ways. First, tracing forms of public participation and the range of meanings embedded in everyday talk provides descriptive accounts that offer practical understandings that could be used, tweaked, and transformed in future public participation design. For example, future public votes should have either a clear advisory or confirmatory function and establish parameters for when another vote is needed to minimize using a call for public vote as an endless way of arguing against an issue. Advisory voting could be used to expand opportunities for thin participation that involve a higher portion of residents. Yet leveraging voting as a preferred form of thin participation may be undermined by the available understanding of voting as a conventional form of public participation, particularly if community members argue that council disagreement with an advisory vote perpetuates distrust, even if the vote is framed as an advisory action. These existing cultural discourses suggest that alternative, new forms of thin participation might create more robust participatory forms with a wider variety of roles for the public rather than focusing on expanding voting.

A cultural approach provides a ground-up method for identifying situated problems within public participation practices—emergent tensions and constraints for participation. For example, working groups prioritize expertise as the primary qualification for entry, which presents democratic contradictions for inclusivity. Drawing on local expertise is more democratic than paying outside consultants, yet this privileged form of participation is extremely exclusive, reinforcing technocratic tendencies to treat public issues as technical issues wherein expertise is foregrounded rather than public values. Future public participation designs could challenge this problem by removing working groups. But our analysis suggests that this practice could also be used to redress the current problem of exclusion by reconfiguring norms for task force selection and the issues they address. A Latinx working group on just transition, for example, could build on this thick participation opportunity to expand what counts as expertise and valued ways of knowing about energy within Boulder.

Second, existing cultural clashes between different norms and expectations for participation can emerge in local discourse. For example, county residents living outside Boulder expect to have the right to vote on issues that will affect their energy supply—even though they are not Boulder residents. Access to public participation should be determined based on the effects of decisions, not on geographic boundaries.

Finally, an orientation to cultural discourses highlights local norms and, in turn, necessitates that design interventions are accountable to community-specific means and meanings. In other words, cultural discourses become inventional resources when these understandings inform public participation designs that reflect local expectations and notions of participation. For example, whereas public hearings and public comment sessions limit two-way interaction, they are simultaneously recognized as valued forms of participation for both council members and members of the public. Innovative public participation practices might seek to add opportunities for thicker participation, public deliberation, and discussion without entirely reducing opportunities for holding officials accountable through conventional participation.

PRACTICAL APPLICATION FOR COMMUNITY ENGAGEMENT INNOVATION

Mapping forms of public participation became a resource for innovation in our own practitioner work. In a larger collaboration with the city of Boulder, city staff complained that community members got stuck telling the city what to do to improve public engagement rather than recognizing their own roles in public engagement. We recognized that one reason for this is that people were focusing solely on city council meetings and blaming the city for their current structure. We expanded the list of culturally relevant forms and commissioned a graphic recording to represent Boulder's broader civic communication landscape. Whereas the analysis presented in this chapter focuses on city-led spaces, the landscape maps the intersections of community-led and media participatory spaces. Community-led spaces included neighborhood associations, advocacy groups, houses of worship, community-organized forums, and conversations with friends, neighbors, and colleagues. Mapped as part of the landscape of mediated spaces were the city's website, Twitter, Channel 8 (the local access cable television channel),

YouTube, Nextdoor, Boulder's *Daily Camera* newspaper, local radio and podcasts, and other community media. The graphic was used in discussions with community members to consider various norms for participation that are active within and across forms. The graphic moved conversation beyond public meetings (i.e., city council public comments)—a form that theorists have long been frustrated with as well—to consider the range of forms available. In this way, the graphic allowed the community to see its own participation differently, enabling community members to reimagine what public participation could look like. Our basic research on cultural forms of public participation (including very conventional forms) was leveraged to enable the community to envision more satisfying community engagement.

Implications for Energy System Transition and Energy Democracy

Our study examines complexities of public participation in energy system transformation. Boulder's discourse suggests that public participation during energy transition is necessary and supported, yet current practices are constrained by institutionalized rules for engagement, expectations for interaction between the council and public, and assumptions about the sorts of expert credibility and technical knowledge necessary for the public to be "informed" decision makers.

Energy system transition requires cultural practices and processes; leveraging public support for changing energy production and consumption hinges on arguments situated in community identity, shared value systems, and situated norms for communication, which, in turn, shape participatory practices. This analysis shows residents of Boulder and Boulder County invoking *expectations* of energy democracy—they claim a democratic right to vote and shape the energy system. These expectations are notable because, in many ways, demands for energy democracy precede scholarship on energy democracy. In Boulder, expectations for energy democracy seem tied to democratic expectations that residents have of city government—even when they live outside the city boundaries. Theorists would benefit from reflecting on the benefits and limitations of drawing on existing democratic norms for city governance as the basis for thinking about energy democracy. Cities offer comparative democratic advantages to corporate utilities, yet Boulder citizens also press for ways to recognize a wider range of stakeholders beyond just city residents. This reinforces the need to recognize a wider range of actors and subjectivities within energy democracy beyond citizen or resident.

Energy system transition is a necessarily technical process, which can lead to a focus on narrow types of expertise that seem essential to understanding energy options. The Boulder case demonstrates an inability to orient to and recognize diverse ways of knowing. Outreach and education discourses draw on deficit model thinking wherein the public lacks knowledge necessary to make decisions about technological issues. Theorizing energy democracy must contend with the challenges of expertise within democracy, such as how to integrate experts to support well-informed viewpoints without crowding out or silencing voices of citizens and other ways of knowing (Sprain et al., 2014). In Boulder, one way this could be done is by expanding the composition of working groups and the range of issues they tackle so that they no longer function as technical consultants. Thick participation opportunities like deliberation can also constitute new relationships between citizens and experts (Sprain et al., 2014).

Energy system transition requires coordinated action between publics, government institutions, and energy institutions over time. Transformation is an ongoing process; our data show six years of this still ongoing process. Boulder recognizes that achieving their climate action goals will only be actualized if changes occur across the system—individuals have to modify personal consumption habits, while the city needs to take larger actions such as changing the electric utility. Transforming an entire energy system entails processes that change material infrastructure, public policies, environmental impacts, and community understandings. Public participation is necessary at all phases of this development and transition. Some of the practices here, such as translating a close vote into a clear mandate and community support while ignoring 49% of voters, have the potential to undermine adaptive governance and capacity to transform over time. In turn, ongoing governance raises concerns about politicizing participation over time (Sprain, 2016) as people develop expertise through their ongoing participation such that they no longer see themselves as representing the public (as we heard in the public comments). Energy system governance calls for practices that help address these inherent tensions within transition.

Conclusion

This chapter explores the role of the public in energy governance by orienting to local practices and meanings for participation. By taking a cultural perspective, we consider the ways in which cultural discourses configure participatory practices and extend public participation theorizing to consider how cultural understandings might be leveraged in designs that better engage the public, enabling decision making and facilitating coordinated actions between the government and citizens that are necessary to make Boulder's energy future a reality. Through our attention to the complexities of participation in energy transitions, we propose that local understandings become resources for designing processes that are responsive to the cultural milieu, extend current processes, and address the unique complexities of public participation in energy system transformation.

References

Araújo, K. (2014). The emerging field of energy transitions: Progress, challenges, and opportunities. *Energy Research & Social Science, 3*, 112–121. doi:10.1016.j.erss.2014.03.002

Bolton, R. & Foxon, T. J. (2015). Infrastructure transformation as a socio-technical process—Implications for the governance of energy distribution networks in the UK. *Technological Forecasting & Social Change, 90*, 538–550. doi:10.1016/j.techfore.2014.02.017

Boulder Weekly Staff (2011, November 1). Boulder municipalization (2B and 2C). *Boulder Weekly.* Retrieved from http://www.boulderweekly.com/article-6823-boulder-municipalization-(2b-and-2c).html

Bulkeley, H., & Mol, A. P. (2003). Participation and environmental governance: Consensus, ambivalence and debate. *Environmental Values, 12*(2), 143–154. doi:10.3197/096327103129341261

Burness, A. (2017a, April 17). No pause; Boulder City Council votes to continue litigation on municipalization. *DailyCamera.* Retrieved from http://www.daily camera.com/news/boulder/ci_30928489/no-pause-boulder-city-council-votes-continue-litigation?source=pkg

Burness, A. (2017b, November 7). Boulder's muni lives as voters reaffirm support for local electric utility. *Daily Camera.* Retrieved from http://www.dailycamera.com/boulder-election-news/ci_31437049

Buttny, R., & Cohen, J. R. (2015). Public meeting discourse. In K. Tracy, C. Illie, & T. Sandel (Eds.), *The International Encyclopedia of Language and Social Interaction* (pp. 1242–1252). Boston, MA: John Wiley & Sons.

Carbaugh, D. (2007). Cultural discourse analysis: Communication practices and intercultural encounters. *Journal of Intercultural Communication Research, 36*, 167–182. doi:10.1080/17475750701737090

Carbaugh, D., & Cerulli, T. (2013). Cultural discourses of dwelling: Investigating environmental communication as a place-based practice. *Environmental*

Communication: A Journal of Nature and Culture, 7(1), 4–23. doi:10.1080/17524032.2012.749296

Carcasson, M. & Sprain, L. (2016). Beyond problem solving: Re-conceptualizing the work of public deliberation as deliberative inquiry. Communication Theory, 26, 41–63. doi:10.1111/comt.12055

Chilvers, J., & Longhurst, N. (2016). Participation in transition(s): Reconceiving public engagement in energy transitions as co-produced, emergent, and diverse. Journal of Environmental Policy & Planning, 18(5). doi:10.1080/1523908X.2015.1110483

City of Boulder, Colorado. (2016). Energy Future: Working Groups. Retrieved from https://bouldercolorado.gov/energy-future/energy-working-groups

Dean, R. J. (2016). Beyond radicalism and resignation: The competing logics for public participation in policy decisions. Policy & Politics, 45(2), 213–230.

Depoe, S. P., & Delicath, J. W. (2004). Introduction. In S. Depoe, J. W. Delicath, & M. F. A. Elsenbeer (Eds.), Communication and public participation in environmental decision making (pp. 1–12). Albany: State University of New York Press.

Endres, D. E., Cozen, B., Barnett, J. T., O'Byrne, M., & Peterson, T. R. (2016). Communicating energy in a climate (of) crisis. Communication Yearbook, 40, 419–447. doi:10.1080/23808985.2015.11735267

Fung, A. (2006). Varieties of participation in complex governance. Public Administration Review, 66(s1), 66–75. doi:10.1111/j.1540-6210.2006.00667.x

Goldthau, A. (2014). Rethinking the governance of energy infrastructure: Scale, decentralization and polycentrism. Energy Research & Social Science, 1, 134–140. doi:10.1016/j.erss.2014.02.009

Hymes, D. (1974). Foundations in sociolinguistics. Philadelphia: University of Pennsylvania Press.

Lassen, I., Horsbøl, A., Bonnen, K., & Pedersen, A. G. J. (2011). Climate change discourses and citizen participation: A case study of the discursive construction of citizenship in two public events. Environmental Communication: A Journal of Nature and Culture, 5(4), 411–427. doi:10.1080/17524032.2011.610809

Martin, T. (2007). Muting the voice of the local in the age of the global: How communication practices compromised public participation in India's allain dunhangan environmental impact assessment. Environmental Communication, 1(2), 171–193. doi:10.1080/17524030701642595

Morgan, E. L. (2003). Discourses of water: A framework for the study of environmental communication. Applied Environmental Education & Communication, 2(3), 153–159. doi:10.1080/15330150390218261

Nabatchi, T., & Leighninger, M. (2015). Public participation for 21st century democracy. Hoboken, NJ: John Wiley & Sons.

Newig, J., & Fritsch, O. (2009). Environmental governance: Participatory, multilevel, and effective? Environmental policy and governance, 19(3), 197–214. doi:10.1002/eet.509

Philipsen, G. (1992). *Speaking culturally: Explorations in social communication*. Albany: State University of New York Press.

Reed, M. S. (2008). Stakeholder participation for environmental management: A literature review. *Biological Conservation, 141*, 2417–2431. doi:10.1016/j.bio con.2008.07.014

Schwarze, S. (2006). Environmental melodrama. *Quarterly Journal of Speech, 92*, 239–261. doi:10.1080/00335630600938609

Scollo, M. (2011). Cultural approaches to discourse analysis: A theoretical and methodological conversation with special focus on Donal Carbaugh's Cultural Discourse Analysis. *Journal of Multicultural Discourses, 6*(1), 1–32. doi:10.10 80/17447143.2010.536550

Senecah, S. L. (2004). The trinity of voice: The role of practical theory in planning and evaluating the effectiveness of environmental participatory processes. In S. Depoe, J. W. Delicath, & M. F. A. Elsenbeer (Eds.), *Communication and public participation in environmental decision making* (pp. 13–34). Albany: State University of New York Press.

Senecah, S. L. (2007). Impetus, mission, and future of the Environmental Communication commission/division: Are we still on track? Were we ever? *Environmental Communication, 1*(1), 21–33. doi:10.1080/17524030701334045

Sprain, L. (2006). Literature review for "Painting the Landscape: A cross-cultural exploration of public-government decision making." Report prepared for an IAP2-Kettering Research Project.

Sprain, L. (2014). Voices of organic consumption: Understanding organic consumption as political action. In J. Peeples & S. Depoe (Eds.), *Voice and Environment* (pp. 127–147). New York: Palgrave.

Sprain, L. (2016). Paradoxes of public participation in climate change governance. *The Good Society, 25*(1), 62–80.

Sprain, L., & Boromisza-Habashi, D. (2013). The ethnographer of communication at the table: Building cultural competence, designing strategic action. *Journal of Applied Communication Research, 41*, 181–187. doi:10.1080/00909882.2 013.7822418

Sprain, L., & Carcasson, M. (2013). Democratic engagement through passionate impartiality. *Tamara Journal for Critical Organizational Inquiry, 11*, 13–26.

Sprain, L., Carcasson, M., & Merolla, A. J. (2014). Utilizing "on tap" experts in deliberative forums: Implications for design. *Journal of Applied Communication Research, 42*(2), 150–167. doi:10.1080/00909882.2013.859292

Sprain, L., van Over, B., & Morgan, E. L. (2016). Divergent meanings of community: Ethnographies of communication in water governance. In T. R. Peterson, H. Bergeå, A. Feldpausch-Parker, & K. Raitio (Eds.), *Environmental communication and community: Constructive and destructive dynamics of social transformation* (pp. 249–266). New York: Routledge.

Tracy, S. (2010). Qualitative quality: Eight "big-tent" criteria for excellent qualitative research. *Qualitative Inquiry, 16*, 837–851. doi:10.1177/1077800410383121

Walker, G. B. (2004). The roadless initiative as national policy: Is public participation an oxymoron? In S. Depoe, J. W. Delicath, & M. F. A. Elsenbeer (Eds.), *Communication and public participation in environmental decision making* (pp. 113–136). Albany: State University of New York Press.

Walker, G. B. (2007). Public participation as participatory communication in environmental policy decision-making: From concepts to structured conversations. *Environmental Communication: A Journal of Nature and Culture, 1*(1), 99–110. doi:10.1080/17524030701334342

Wills, Toker, C. (2004). Public participation or stakeholder frustration: An analysis of consensus-based participation in Georgia Ports Authority's stakeholder evaluation group. In S. Depoe, J. W. Delicath, & M. F. A. Elsenbeer (Eds.), *Communication and public participation in environmental decision making* (pp. 175–200). Albany: State University of New York Press.

Chapter 6

The Radical Potential of Public Participation Processes

Using Indecorous Voice and Resistance to Expand the Scope of Public Participation

Kathleen P. Hunt, Nicholas P. Paliewicz, and Danielle Endres

Most people who showed up never got a chance to testify. Oh, yeah, you can go in the closet with the woman that's, that's taking down the testimony. That's not a hearing. A hearing is when you get heard. A hearing is when you are heard. When you get to hear the other people. That's why they come out. We all know that we can send in comments to Carol on e-mail. By snail mail. By whatever means, fax. We can do that. We know that. But you come to a hearing in order to be heard. And most of the people that came here were not heard.

—Judy Treichel, executive director of the
Nevada Nuclear Waste Task Force (Treichel, 2001)

Public participation processes, including meetings, comment sessions, hearings, and information sessions, bring a group of people together in the same space-time who want to *be heard*. Instead of emailing or mailing in a comment, those who show up to these public events do so because of the unique rhetorical opportunities they present, including, for example, interactions with elected officials and key decision makers, and engagement with other participants. Thus, these rhetorical encounters involve a variety of perceived audiences and rhetors: the officials tasked with running the event, participants seeking to be heard, and others (such as media or other publics) there to hear the participants. Within this space, we can see the multidirectional nature of public participation, with varied interactions and multiple audiences revealing the complexity of rhetoric at play in these processes.

We suggest that public participation processes, though often dismissed as forms of Decide, Announce, Defend (DAD) decision making that limit voices of dissent (e.g., Beierle & Cayford, 2002; Bucchi & Neresini, 2007; Depoe, Delicath, & Elsenbeer 2004; Dietz & Stern, 2008; Endres, 2009, 2012; Hendry, 2004; Kinsella, 2004; Senecah, 2004; Walker, 2007), are also important moments of radical rhetorical engagement and consequence. Acts such as obstructing the flow of a meeting by yelling or clapping, or openly defying rules such as length of time for oral comments, violate normative rules of decorum, or the expectations of proper behavior, that are set up by formalized public participation processes. Yet these forms of indecorous behavior can also contribute to solidarity and community among attendees, enact image events that garner media attention to what might otherwise be seen as a routine and non-newsworthy event, and offer opportunities for participants to create their own media content. These moments, often excluded from official transcripts and, by extension, academic scholarship, should not be dismissed as violations of decorum, but rather should be examined as moments that reveal the complexity and opportunities inherent in even the most restricted DAD models of public participation.

On January 10, 2007, about 100 people gathered at the Grand America Hotel in Salt Lake City for a "public information session" hosted by the Department of Defense's (DOD) Defense Threat Reduction Agency (DTRA) and the Department of Energy's (DOE) National Nuclear Security Administration (NNSA). This session was devoted to information about the "Divine Strake" test—a non-nuclear high-yield test of a 700-ton buried chemical explosion—that was proposed for detonation in the Nevada Test Site (NTS; now Nevada National Security Site), where more than 1,000 nuclear bombs had been tested from 1951 to 1992. The Divine Strake test would contribute to ongoing military research on bunker-buster weapons. Given the wounds from decades of nuclear testing among Utah downwinders, information about this public information session quickly circulated among activists and citizens concerned that the Divine Strake test, although an underground non-nuclear test, would uproot radioactive dust from the NTS or was a sign of an impending return to nuclear testing. Although there were court reporters available to take individual statements from attendees, there was no opportunity for attendees to speak to or be heard by the whole forum that had gathered that evening. Rather, participants mulled around the hotel ballroom, which contained a variety of informational posters about the scientific viability and safety of the Divine Strake test. The layout of the room, the diffused posters that encouraged one-on-one interactions between the DOD/DOE employees and attendees, the conversations and questions that were not recorded, and the lack of podium and microphone signaled to attendees that they were meant to learn about the Divine Strake test but not to officially offer their comments. From one perspective, this session was a classic example of a DAD model of public participation; from another perspective, this public information session offered an opportunity for resistance, an opportunity that was taken up by several attendees.

In the midst of the session, Kevin Donahue yelled out: "Who in this room is against Divine Strake?" followed by a resounding "We are!" (Dickson, 2007, n.p.).[1] This immediately attracted everyone's attention, including that of the local television news crew. Plainclothed security guards quickly began to escort the man from the room. As he was grabbed, the man called out: "I thought this was a public meeting," to which one of the guards replied, "This is NOT a public forum" (Dickson, 2007, n.p.). As the man was escorted from the room for breaking the rules of the session, his indecorous outburst seemed to express a collective sense of frustration that had been gathering in the room among attendees. Indeed, this "disruption" was what everyone talked about for the rest of the event and beyond; it became the lead in media coverage (Bauman, 2007; Fahys, 2007). It was used as an example to point out the flaws in the public information session, and it rallied both supporters and dissenters of the Divine Strake test.

Donahue's actions constituted a significant moment of disruption in the planned public information session and had important effects. The day after the public information session, the governor of Utah expressed his frustration and announced that he would hold two public hearings so that public comments could be included in the state's response to the Divine Strake proposal. Furthermore, Utah Senator Orrin Hatch and Representative Jim Matheson called on the Pentagon to offer a genuine opportunity the public to voice their opinions and ask questions about the Divine Strake test ("Divine Strake," 2007; "Governor sets," 2007).

This moment of resistance should not be written off as just another activist making trouble. Rather, it reveals the radical potential of public participation processes. In this moment, the traditional model of bidirectional interaction between the public participant speaking to or hearing from an official decision maker broke down in favor of an expanded notion of the public hearing as a multidirectional interaction among participants—Kevin Donohue speaking to the other attendees—and between these participants and wider publics through the media spectacle that ensued. Moreover, this moment calls attention to the ways activists might use public participation processes for their own rhetorical purposes that are untethered to the official agenda. With this in mind, we seek to expand notions of what can count as public participation by shifting focus from official accounts (i.e., transcripts, comment summaries, and other regulatory documents that result from a public session) to the complex, varied, and multidirectional rhetorical interactions that can happen in these settings and may be left out of official summaries. Such a shift in focus requires moving beyond an assumed boundary between activism and public participation by expanding our conception of the rhetors, audiences, purposes, and consequences of public participation processes.

We begin our investigation into the radical potential of public participation processes with an examination of the scholarship on public participation in environmental decision making and its (lack of) intersection with social movement scholarship. From this we offer a critique of the dominant model of public participation, arguing that violations of decorum expose the multiplicity of rhetorical encounters that can take place in these sessions. Analysis of public meetings concerning toxic waste in Love Canal and a public hearing regarding air pollution in Salt Lake City reveals moments of resistance, disruptions, and violations of decorum that demonstrate the ways that individuals and activist groups can use public participation processes to cultivate identity with their cause, enact media events, and ultimately reach new audiences. We conclude with implications of these findings for expanding scholarship in public participation in environmental decision making through productive intersections between this area and social movement scholarship, and highlight the rhetorical consequences of indecorous voice.

Indecorous Voice:
Binding Public Participation and Social Movement

Public participation is a significant area of study within environmental communication. As Cox (2013) defines it, public participation in the United States is "the ability of individual citizens and groups to influence decisions through (a) the right to know or access relevant information, (b) the right to comment to the agency that is responsible for a decision, and (c) the right, through the courts, to hold public agencies and business accountable for their environmental decisions and behaviors" (p. 84). This definition, and much of the research in the field, assumes that public participation in environmental decision making is an *official process* through which citizens have a voice in the environmental decisions that affect and matter to them. By official processes, we mean public hearings, public information sessions, public comment periods, and other modes designed by decision makers to gather input from affected stakeholders and publics. These processes are most often initiated by government entities and defined in particular policies, such as NEPA, and through regulatory agencies (EPA, NRC, DOE, etc.). Research in public participation in environmental decision making is largely focused on either exposing flaws in current models that limit public involvement (e.g., Davies & Selin, 2012; Depoe et al., 2004; Endres, 2009; Katz & Miller, 1996; Philips, Carvalho, & Doyle, 2012) or proposing new models that would better enact the rights listed in the definition above (Beierle & Cayford, 2002; Callister, 2013; Daniels & Walker, 2001; Fiorino, 1990; Fischer, 2000; Hamilton & Wills-Toker, 2006; Kinsella, 2004; Senecah, 2004).

In this chapter, we push beyond the study of the limitations of and opportunities for better public participation processes. Current scholarship would benefit from expanding the definition of public participation from an official process that entails particular rights and responsibilities to a definition that recognizes the multiplicity of rhetorical interactions that can happen within these processes, many of which challenge the official purpose of public participation. Pezzullo (2007) and Delicath (2004) have called on public participation scholarship to expand its notion of what public participation can be to include activism and other forms of action that occur outside official processes, such as protest rallies, art installations, and toxic tours. We agree with these important challenges, yet take up the impulse to break down barriers between activism and public participation in a different way. Instead of looking outside official public participation processes for alternate modes of participation, we argue that alternative, resistive, and transgressive rhetorical practices happen *within* public hearings, public information sessions, and other forms of sanctioned participation. In other words, the multidirectional nature of public participation processes presents a variety of opportunities for rhetorical engagement.

Traditional public participation scholarship assumes that public hearings and other processes of public participation are an instrumental form of communication wherein participants speak directly to decision makers in an attempt to influence their decisions about an impending issue of importance. Drawing from Randall Lake's (1983) distinction between instrumental and consummatory rhetoric, we use instrumental to mean rhetoric that is a means to a particular end, in this case, seeking institutional change within official public participation channels (see also Bowers, Ochs, Jensen, & Schulz, 2009; Cherwitz & Zagacki, 1986). We do not deny that the instrumental model is one way of understanding public participation processes. However, the almost exclusive focus on this understanding has limited consideration of other rhetorical dynamics at play in these contexts of environmental decision making.

As the examples that open this chapter demonstrate, participants may *also* hear and speak to *other* participants and not only to decision makers attending these events. Sometimes, whether planned or spontaneous, participants may violate expectations or norms of decorum and disrupt the official process, creating media events that disseminate to other audiences beyond the public hearing. In this way, participants may engage in forms of consummatory rhetoric, an end in itself (Lake, 1983), such as voicing an opinion, being heard, or constituting an identity (Charland, 1987). This complex mix of rhetorical opportunities comes into sharper focus when we move beyond a strictly instrumental definition and consider the radical potential of public participation processes.

Specifically, we are interested in how activism, which we define as tactics of resistance designed to bring about social change (de Certeau, 1984), can happen within public participation processes. Moments of disruption and violations of decorum during a hearing or meeting may or may not influence the specific decision-making process, yet they can contribute to a movement of meaning over time (McGee, 1980). For instance, a disruptive event might introduce a new perspective, or mind-bomb, into the consciousness of other attendees. Or an image event created by public hearing attendees might raise collective public awareness about an environmental issue and eventually, when combined with other moments, change the way people understand their relationship with the environment.

ARTICULATING PUBLIC PARTICIPATION AND SOCIAL MOVEMENTS

Environmental communication has made invaluable contributions to the study of both social movement and public participation. Yet these threads of scholarship are more frayed than interwoven, contributing to an unnecessary bifurcation of traditional and alternative modes of public engagement. Grounded in the work of Cox (1982) and Oravec (1984), one area of our scholarship has taken up critical analysis of rhetorical tactics used in environmental movements (e.g., DeLuca, 1999; Endres & Senda-Cook, 2011; Hunt, 2014; Pezzullo, 2007). Another area has focused on modes and methods of institutional public participation (Callister, 2013; Endres, 2009; Senecah, 2004). Indeed, the field's first volume on public participation (Depoe et al., 2004) cordons off case studies of environmental activism under the heading "Emergent Participation Practices Among Activist Communities." Constituting public hearings as official processes of participation, and protest events as alternative (or unofficial) forms of participation, treats both as independent rhetorical situations with discrete purposes, audiences, and rhetors.

Such moves, though unintentional, can privilege the institutional discourses of official public participation processes even while criticizing them as antidemocratic (Cox, 1999). Viewing activists such as Kevin Donohue, who was removed from the Divine Strake public information session, as a disruption reinforces the notion that public participation is a rational, civil, and normative form of discourse and that activism is an alternate form that does not fit within official, instrumental models. If institutional public participation processes are viewed as the norm, and activism as an alternative (and, thereby, separate) process, we risk further marginalizing, and even silencing, those who experience and express the effects of environmental injustice firsthand. In so doing, we neglect consideration of the complexity and radical potential of public participation processes, including creative forms of dissent, multiple means of engagement, and the interplay between internal and external audiences attending (to) such processes. As a corrective, we entreat environmental communication scholars to consider the ways in which public participation, social movement, and activism are entwined.

Scholarship on the rhetoric of social movements reveals the contingent boundary between public and private, values and practices, institutions and people that protests, rallies, and marches can make visible. Social movement rhetoric "violate[s] the proposition that, in an orderly society, there must be prescribed times, places, and manners of protest" (Haiman, 1967, p. 15). Indeed, this area of scholarship has catalogued an array of "non-rational" modes of discourse as tactics of protest including sit-ins and self-immolation (Haiman, 1967) and even the use of profanity and obscenity (Windt, Jr., 1972). Such work has necessarily expanded our understanding of what counts as rhetoric, where protest happens, and to whom these tactics are aimed (Campbell, 1973; Enck-Wanzer, 2006; Lake, 1983). While the assumption that activism is primarily a tactic of resistance that disrupts and reveals alternatives to the status quo (de Certeau, 1984) pervades this literature, it should not imply a simplistic inside/outside dichotomy in terms of official processes of democratic participation—that is, that all activists and modes of protest exist outside traditional discourse and ordered arenas. As Haiman (1967) suggests, "permissible time, place, and manner [of protest]" is only a "proposition," often reinforced by scholars (p. 14).

Activist tactics serve both internal and external audiences and purposes (Cathcart, 1978; Chávez, 2011; Endres, 2011; Gregg, 1971; Windt, Jr., 1972). This is significant in its recognition that social movement rhetoric is not always directed at decision makers, including those convening a public hearing. Activists attending a public meeting may be more interested in rhetoric that coalesces attendees, raises awareness among other participants in the meeting, creates a media event that can reach new audiences, and/ or performs resistance as a consummatory act. Indeed, contemporary social movement scholarship reveals that activists seize a variety of opportunities for engagement that range from working within institutional channels to disrupting those channels.

Indecorous Voice as the Tie That Binds

Indecorous voice—rhetorical tactics that are perceived to be disruptions of proper behavior—is a useful heuristic for engaging the intersection of social movement and public participation. Indecorum is often levied as a pejorative label for improper public participation, as in instances of public agencies dismissing participants as hysterical (Cox, 1999) or "useless" housewives (Gibbs, 2011). We suggest that indecorous voice may not (only) be what de Certeau (1984) calls a strategy or a "structural . . . construction" of rational discourse (Cox, 1999), lobbied against unruly publics by officials, but may be an activist tactic taken up by participants as a form of engagement itself. Tactics of indecorum rhetorically question the very conditions of what counts as appropriate participation, call attention to how instrumental models (such as DAD) can stymie oppositional voices, and reveal the multidirectional nature of public participation processes. Indecorum creatively widens the scope of participation tactics and audiences; transgressing rules of decorum through acts of resistance can serve consummatory functions, including engendering shared identities of protest (Gregg, 1971) and being heard by multiple audiences.

Rooted in Ciceronian (1961) concepts of moral goodness, decorum references a form of propriety to which rhetors are held accountable when speaking in various contexts. In public participation processes, decorum articulates an appropriate discursive style, or normative rules for rational engagement between publics and institutions, embodying the assumption of bidirectional interaction within asymmetrical power dynamics between publics and decision makers (Hariman, 1992). Public hearings can, for instance, restrict the number of testimonies, impose time limits for comments, confine statements to selected topics (such as scientific and technical evidence), and exclude "non-rational" forms of discourse (such as clapping, yelling, or cheering) from official transcripts (Endres, 2009; Hunt & Paliewicz, 2013). In these moments, actors' indecorous voice can be dismissed on the basis that their standing is "inappropriate" or not fitting for the context (Cox, 1999).

When used as a tactic, *in*decorum can demonstrate the radical potential of public participation processes. By radical, we mean both a departure from what is ordinary or traditional, as well as more extreme forms of resistance that challenge institutionally driven public participation processes. Indecorous tactics such as chanting in unity, clapping or booing, crying, or sitting in silence disrupt "proper" participation processes, transgress the "appropriate" discursive style of rational public engagement, and challenge normative rules of "order." As noted above, indecorous voice can result in image events (DeLuca, 1999), spatial events (Endres, Senda-Cook, & Cozen, 2014), or other mind-bombs that radically challenge and attempt to rearticulate dominant grids of intelligibility. In these ways, indecorum can be deployed toward other ends beyond the official purpose of the public hearing or meeting that can forge new alliances with a variety of audiences.

Enacting Indecorous Voice in Public Participation Processes

Environmental activists can engage in public participation processes to disrupt, resist, and make their voices heard. Using tactics alternative to instrumentalist forms of participation, publics can defy the expectations of proper behavior to express communal frustration or anger, enact solidarity with a shared cause, garner attention from external audiences (such as the media), and expose injustices of the public participation process itself. Our analysis hones in on two case studies in which environmental activists enacted indecorous voice for rhetorical consequence: Love Canal and Rio Tinto Kennecott. From 1976 to 1981, residents of Love Canal cried, screamed and hollered, and chanted during public sessions to call attention to the health impacts of chemical toxicity in their community. More than thirty years later, activists in Salt Lake City raised signs, cheered, and voiced their opposition during a Department of Environmental Quality (DEQ) public hearing on air quality. Spanning a 30-year arc, these cases demonstrate the multidirectional nature of public participation and the range of indecorous tactics activists can use to disrupt institutional norms of decorum, be heard by multiple audiences, and enact resistance within institutional public participation processes.

Love Canal Residents Fight to be Relocated

Love Canal is a central case study of environmental (in)justice in the United States (Blum, 2008; Jamieson, 2007). Through grassroots organizing, Lois Gibbs and the Love Canal Homeowners Association (LCHA) pressured for state and federal investigations and, eventually, an injunction to relocate residents affected by toxic waste. Love Canal remains a historic model of grassroots environmental activism, illustrating the radical potential of public participation processes and the consummatory function (Lake, 1983) of indecorum. Through their tactical enactments of indecorous voice, Love Canal residents came to identify as an affected group, constituting a community of affected homeowners while also propelling their local struggle to national prominence via extensive media attention.

Located in upstate New York, Love Canal is a small working class community built over a chemical plume in the early 20th century. Residents first became aware of chemical toxicity when a local newspaper published reports in 1976. Over the next five years,[2] residents would endure innumerable public hearings and various meetings with New York State legislative and health officials. Promises of evacuations were followed by government backtracking as parties debated responsibility. Throughout the unfolding of these events, authorities carefully managed decorum within a traditional DAD model of public participation, limiting Love Canal residents' access to relevant information and marginalizing local concerns for community health and safety. For example, when residents presented evidence of increased infant mortality and other troubling health impacts, the EPA refuted this as informal evidence and "useless housewife data" and, when pressed, arranged for the residents to be examined by a veterinarian (Gibbs, 2011; Matthews, 1997). It took years of public meetings, hearings, and protests[3] for President Jimmy Carter to declare Love Canal a national emergency, approving the final relocation of all remaining residents on October 1, 1980.

Challenging the formal nature of the public participation process, locals vigorously contested institutional procedures and normative expectations for public engagement through the tactical use of indecorum. Openly transgressing rules of rationality and order, residents would be "screaming and hollering," crying, and shouting (Gibbs, 2011, p. 56) during public meetings and hearings, creating an atmosphere that was described as "not pleasant" (Levine, p. 48), "intense," and even "chaotic" (Gibbs, 2011). Even when asked by an official to "calm these people down," Gibbs demurely suggested that community members "sit quietly and listen to the questions and listen to the answers," but quickly added a defiant call to "boo the hell out of them" (Levine, 1982, p. 36). The official's request for calm illustrates the "structural" source of indecorous voice (Cox, 1999), the grounds from which boisterous engagement is dismissed as disruptive. Through her bold entreaty, community members take back the pejorative construction, using indecorum to rudely voice their opposition. Indeed, Gibbs and the residents learned to exploit the very emotionality typically dismissed under the rubric of decorum: "We used to use Patty . . . who was one of our best criers, who would stand up and shout . . . and cry," often while also holding her infant child at public meetings (Matthews, 1997). Indecorum is thus a deliberate activist tactic, as Love Canal activists ". . . thought about how people behaved and how could we use that to heat up the struggle and put pressure on those target people" (Matthews, 1997).

Residents' chanting and shouting, Gibbs (2011) reflects, often unfolded "like an opera or a musical, with those up front questioning the commissioners like lead singers, and . . . the audience would cry out just as if they were the chorus" (p. 154). This comment illustrates how participation in public processes often extends beyond simple one-way interaction between attendees and officials; those "in the chorus" use their indecorous voices to support and amplify other "singers." For example, documentary footage of one public hearing shows a middle-aged man standing up with arms outstretched in exasperation, shouting: "All I want, all I want . . . is my 28-5 [referring to the government payment for his home] and give it to me tonight and I'll move down that road and I'll never look back at the Love Canal *again!*" On the verge of tears, his face is flushed and his voice trembles. Immediately following his statement, the crowd erupts in boisterous cheers. Community members clap and chant, "WE WANT OUT! WE WANT OUT!" in unison, while others can be seen pumping their fists and jumping from their seats (Matthews, 1997). In this moment, the angry resident is not only expressing frustration to officials, but also is igniting the similar feelings of others, who then are compelled to join in as a chorus.

Indecorum thus allowed the affected residents to speak not only to the decision makers, but also to one another.

The "strong elements of confrontation" displayed by Love Canal residents during public participation processes "provided emotional catharsis" (Levine, 1982, p. 23) for them. As such, we submit, indecorous tactics created a shared identity of protest, or what Gregg (1971) refers to as the "ego-function" of social activism. Gibbs (2011) consistently articulates the struggle of "ordinary citizens [gaining] power" as *our community's fight* (emphasis added, p. 2). Indeed, the formation of the LCHA was itself fomented as an effort to "transform an angry crowd" into organized activists (Blum, 2008, p. 27). Public meetings and hearings thus "became the context in which . . . a *group* was forming" as Love Canal residents, who may not have previously known or interacted with one another, "began to learn that they shared common problems . . . [and] began to develop the deep understanding that they would have to depend on themselves and one another" (Levine, 1982, p. 50). Regardless of the outcome of the particular public meeting and whether any decision makers were persuaded, the expressions of pain, anger, and solidarity were heard by more than the officials present and contributed to an internal rhetoric of collectivity. In other words, as residents shared their private plight with chemical toxicity to state legislators and public health officials—painful experiences of miscarriages and stillbirths, disabled children, sick and dying pets—cheers and applause from the crowd function to make "the people [feel] drawn together" (Levine, 1982, p. 58).

Love Canal residents' indecorum was also a tactical move with rhetorical consequences for external audiences, including the media, local and state officials, and the wider public. Media attention became a weapon—indeed, note the Gibbs' description of "targets" above—the residents used to gain exposure, thereby "reach[ing] a public larger than those present" (Levine, 1982, p. 32) and pressuring state and national leaders to act. As residents came to witness multiple attempts to circumvent or deny their complaints, raucously (and sometimes violently, e.g., Matthews, 1997) expressing their frustration at repeated public meetings, "news reporters, photographers, and television cameramen . . . recorded what was happening and inevitably became part of the scene themselves" (Levine, 1982, p. 37). For example, when Governor Hugh Carey sent an assistant in his stead to present an early plan to only evacuate families with pregnant women and young children, Gibbs audaciously shouted directly into a news camera, "Where's Carey? I'd be here if I was governor!" Her veracity incited others nearby to shout with her, "Where's the mayor?" and "Where's Hooker [Chemical Company]?"

(Levine, 1982, p. 37). Such actions clearly defy expectations for rational engagement at a public meeting, illustrating how public attendees do not always solely attend meetings to communicate with officials, and can "make do" (de Certeau, 1984) with what is available to raise collective awareness about environmental (in)justice. Indeed, Gibbs (2011) would later reflect: "We had to keep the media's interest. That was the only way we got anything done. They forced New York to answer questions. They kept Love Canal in the public consciousness" (p. 120). In this way, the "structural source" of indecorous voice (Cox, 1999) became a rhetorical resource for propelling a local struggle into a nationally known event.

Indecorous voice, a term that typically designates inappropriate behavior and normative rules of proper public participation, was tactically reclaimed as Love Canal residents disrupted expectations of bidirectional and rational interaction between the public and decision makers. Although actions like clapping and cheering, chanting, and engaging the media may not necessarily be radical in their singularity, they transgress the appropriate discursive style of public engagement and can be ways of creating a shared identity. The case of Love Canal demonstrates how, through indecorum, activists can seize the multidirectional nature of public participation processes to cathartically voice their resistance to internal and external audiences and contribute to a movement of meaning. Decades later, Salt Lake City activists used indecorum to radically renegotiate the boundaries of participation and protest in a public hearing.

Utah Residents Fight for Clean Air

In 2011, severe air pollution became a political linchpin for activists concerned with tensions between the financial interests of industry and the health effects of poor air in Salt Lake City (Penrod, 2016). This area of Northern Utah experiences some of the worst air quality in the United States (Bennion, 2013), garnering national media attention for regular occurrences of pollution levels above federal standards (Frosch, 2013). Rio Tinto Kennecott (RTK) is one of the largest employers in the Salt Lake Valley, operating the Bingham Canyon Mine, the largest open pit copper mine in the world; mining contributes to nearly one-third of the area's total greenhouse gas emissions (Klaus & Mayhew, 2012).

On February 22, 2011, the Utah Department of Environmental Quality (UDEQ) convened a public hearing to evaluate plans to expand the Bingham Canyon Mine, granting RTK a 30% increase in annual mineral extraction.

Permitting an expansion of this magnitude necessitates a renegotiation of Utah's State Implementation Plan (or SIP, whereby particulate pollution is regulated), for which the Clean Air Act requires an opportunity for public comment. Nearly 150 participants, including environmental activists, unaffiliated members of the public, RTK employees, and elected representatives from local municipalities, attended this event.[4]

The hearing exhibited several hallmarks of DAD decision making, including shorter time limits for oral comments and seating arrangements that strategically placed RTK employees and local officials closer to the UDEQ board. Although the roster of slated commenters was overcrowded and attendees were not instructed on where to sit, institutional arrangements privileged powerful voices and limited opportunity for dissent. In response to these "structural sources" (Cox, 1999) of decorum, Salt Lake activists indecorously shamed the public participation process, with several comments attacking the UDEQ and RTK. For example:

> . . . it appears that DAQ has already decided that this issue will be approved . . . and has produced the completed permit stating just that. Imagine our surprise . . . This situation then makes a mockery of this particular public comment event . . . Obviously these Utah voices . . . are not being considered. (Utah DEQ, 2011a, p. 31).

Comments like this illustrate that, despite recognition of a DAD-style public hearing, participants still found value in being present exposing the absurdity of these UDEQ public hearing conditions. Actions like clapping and booing, holding signs and banners, and sitting in silence tactically and creatively renegotiated rules of decorum, expressing collective frustration, showing solidarity, and galvanizing an oppositional agenda among publics who had not previously interacted.

Before and throughout the hearing, the various grassroots environmental groups in attendance, including the Utah Chapter of the Sierra Club, Utah Physicians for a Healthy Environment, Utah Moms for Clean Air, and WildEarth Guardians, displayed numerous signs as if at a protest rally. One asked, "Is Utah Zion or Mordor?" Another read, "Another Dirty Error From DAQ." Others said, "Got Lungs?" "Protect Our Right to Breath[e]," and a 12-foot banner read, "Rio Tinto: Change We Don't Believe In." Only those in opposition to the mine's expansion brought signs; there were no pro-RTK posters. Although a clear import of an "alternative" protest tactic into an institutional setting, signs also transgress normative rules of decorum in other important ways. Signs can be simultaneously displayed to officials and other audiences (including other attendees as well as media cameras). For example, Beverly Terry's sign, illustrating drops of blood pooled into dollar signs with the message "Our children die so you can make a profit," was featured in the *Salt Lake Tribune*'s coverage of the hearing (Stettler, 2011, para. 12). Furthermore, signs may be held up throughout the hearing so that their message is not limited to the prescribed duration of an oral comment. The content and use of posters and banners (for example, raising or shaking at opportune moments or sharing the task of holding up a banner) contribute to feelings of excitement and solidarity among participants. Indeed, activists held signs extra high when comments favored RTK. This fluidity and rhythm illustrate how protest signs can be used within a public hearing to rebuke pro-industry commenters and contribute to a shared sense of oppositional identity. As a rhetorical tactic of social protest (Gregg, 1971; Haiman, 1967; Windt Jr., 1972), these actions open spaces for thinking differently about how public participation processes ought to occur. By bridging public participation and social protest, Salt Lake City activists alienated RTK from the very process typically structured to serve industry interests.

While signs exceed the boundaries of oral commentary, silence can tactically transgress the assumption that comments must be oral at all. Upon arriving at the front of the room to present his comment, one participant physically moved the chair that, up to this point, had been facing the UDEQ to instead face the other attendees. He asked a simple question and then sat in silence: "For the remainder of my time I invite everyone to consider in silence . . . what is it you love most in this world, and how can you live in integrity with that thing that you love most?" (Utah DEQ, 2011a, p. 44). And the room did indeed sit in silence for several minutes until this participant disclosed that what he loves most is his children. After his remark, the room immediately erupted with applause, cheers, and standing ovations. This moment of silence was a creative consummatory act that brought participants together in a performance that exceeded the norms of bidirectional public participation, inserted a pause in the process, and served to create community between participants.

During this public hearing, participants also used indecorous voice in ways similar to the Love Canal residents. Acts such as clapping, shouting "woo-hoos," and cheers broke from normative expectations of decorum in this space. People built solidarity by collectively reacting to comments that openly opposed the expansion, RTK, and/or the decision-making process itself. For example, one female participant who discussed citizens' rights in a participatory democracy received loud applause and energetic cheers from supporters in the room. Our field notes record an interesting pattern as oral comments proceeded: oppositional comments received boisterous applause and vociferous cheers from activists (and silence from RTK employees and elected officials), while comments supporting the expansion or RTK were met with silence from activists (and mild-mannered applause from pro-industry attendees). Importantly, actions like clapping and paralinguistic devices such as cheering as well as the visual signs and banners are not included in the official documentation of this hearing, but are explicitly mentioned in local media coverage (Stettler, 2011).

Like Love Canal residents, Salt Lake Valley activists indecorously called on other official audiences not in attendance at the hearing to pay attention to air quality. For example, a participant associated with the group Utah Physicians for a Healthy Environment (UPHE) delivered his comment wearing a white medical lab coat, telling an emotional story about a patient of his stricken with a severe respiratory illness. Ending his story with a loud injunction, this speaker directly calls out state officials: "DAQ, legislators, Governor Herbert, Rio Tinto are you *listening*?!" (emphasis added, Utah DEQ, 2011a, p. 37). With this, he received great applause and many cheers. Importantly, the meaning of this moment was carried beyond the walls of the UDEQ meeting room as UPHE and other activist groups used momentum from the hearing to advance other initiatives for clean air (Stettler, 2011). Thus, while indecorous acts serve to expose the flaws of public participation processes in a way that may not persuade the immediate audience of UDEQ decision makers, they can open the possibilities of a multiplicity of audiences and purposes.

Indeed, the Utah Air Quality Board ultimately approved RTK's request to expand the mine, thus permitting more air pollution to affect Salt Lake Valley residents (see Utah DEQ, 2011b). Yet our analysis demonstrates how indecorous voice can be creatively deployed to politicize the instrumentalization of public participation processes. As a rhetoric of social protest, indecorum breaks from what is proper to renegotiate the very conditions of public participation. Participants at the Salt Lake City hearing enacted indecorum to protest a predetermined outcome, make oppositional voices heard, and galvanize solidarity among affected stakeholders.

Conclusion

Public participation, when defined as a bidirectional instrumentalist exchange between publics and decision makers, is untenable to forces of indecorum that contest its very design. This chapter has demonstrated how public participation can become a site of tactical resistance with rhetorical consequences that exceed the specific outcome of the hearing, meeting, or information session. Both Love Canal and Salt Lake City activists, in recognition of the hopelessness of making a difference within a DAD model, use public participation processes as a way to be heard, access different audiences, create media events, and constitute community. Their disruptive acts not only highlight unjust models of participation but also expand beyond the instrumental purpose of the public participation process. In other words, indecorous voice can serve a purpose in itself as a form of consummatory rhetoric (Lake, 1983), including galvanizing a shared identity, being heard, and reaching internal and external audiences.

As case studies of public participation fighting environmental (in) justice, Love Canal and Salt Lake City reveal the interwoven nature of participation and protest, how tactics of indecorum have survived and evolved to meet processes of public participation, and what is likely a tactic used more often than scholars have realized. Residents of Love Canal and the Salt Lake Valley seized approaches such as chanting, holding signs, and engaging the media that often typify social movement (Gregg, 1971; Haiman, 1967; Windt Jr, 1972) and grassroots environmental organizing (Delicath, 2004), within traditional spaces of public participation. To be sure, these "alternative" approaches are often chastised as non-rational, emotional, and uncivil forms of public engagement, and, as such, strategically position discourses of decorum (Cox, 1999) that dismiss them (Cox, 1999). Yet activists can seize this "impropriety" (Cicero, 1961) as a resource for enacting rhetorical consequence; Love Canal residents tactically deployed boisterous behavior to garner media attention, while Salt Lake activists used silence and signage to expose an unjust process. Although actions like shouting, booing, and clapping may not be radical in their singularity, they may work in concert with other tactics and reach beyond the confines of a public meeting room or comment card.

Many of the tactics used at Love Canal were also observed at the Salt Lake City hearing, illustrating how participants continue to "make do" (de Certeau, 1984) in similar ways under comparable circumstances. While both cases protested institutionalized models of public participation processes, new processes structured to more severely limit public involvement and avoid disruption are emerging; publics must continue to adapt indecorous tactics. For instance, the DOD "public information session" on Divine Strake represents a new model of public "engagement" that wholly eschews the expectation for participants' commentary or feedback. Unlike a traditional public hearing that is organized as a venue for public speech—stakeholders offering comments before a panel of decision makers and an audience—the public information session is more diffuse. At the Divine Strake session, posters were displayed for attendees' perusal, encouraging short conversations in small groups with DOD/DOE representatives. Indeed, the security guard who escorted Kevin Donahue out of the room said, "This is not a public forum"; the public can ask questions about scientific posters but not gather as a community. Such information sessions force participants to find new ways of engaging indecorous forms of participation to be heard, voice opposition, and garner media attention.

Despite the long-standing approach of social movements to use indecorum to disrupt and open the radical potential of public hearings, scholarship in public participation in environmental decision making has been slow to focus critical attention on this phenomenon. To be sure, although Love Canal remains a hallmark of environmental justice, the latent assumption (and privileging) of bidirectional communication between public and decision maker has prevented acknowledgment of the radical tactics these activists used and their myriad rhetorical consequences. Amid important research demonstrating flaws in current models and proposing more just processes of participation, rhetorical investigation of the resistive possibilities and radical potential within these processes can reveal how social movements are creatively "making do" (de Certeau, 1984) in an imperfect system. The official summary of the Salt Lake City public hearing only transcribes the oral comments presented before the UDEQ (Utah DEQ, 2011a); the signs and banners and shouts and cheers are not accounted for in agency documentation of this event. Environmental communication scholarship that relies solely on institutional documentation of public participation processes risks neglecting rich and radical rhetorics of protest while also (re)marginalizing voices of dissent.

Our analysis has important implications for future research in public participation and processes of environmental decision making. First, we have revealed the limits of the likely unintended bifurcation between official and alternative processes of public engagement (e.g., protest events) wherein social movement activism is seen as outside the realm of public participation. In addition to highlighting how public hearings involve a range of rhetorical opportunities, audiences, and purposes, this chapter highlights a need for greater engagement at the intersection of public participation and social movement. Building from our analysis of indecorum as one mode of activist engagement within public participation processes, we call for more research that examines other forms of radical engagement.

Finally, our analysis of the use of indecorous voice in public participation processes expands on the concept as first introduced by Cox (1999). While he focused on highlighting how indecorous voice could be a strategy of control wherein decision makers would call out and exclude participants who violate decorum, our focus reveals how indecorous voice can also be a tactic of resistance that can successfully raise awareness about a movement, reach new audiences, and create media events. Taken together, we offer a more complex and robust understanding of the radical potential of indecorum in public participation processes.

Notes

1. Note that one of the authors attended this public hearing and took notes. This section is based both on her recollection of the event and on an article (Dickson, 2007) that appeared in a local magazine about the event.

2. For a complete timeline of these historical events, see the Love Canal archives webpage: http://library.buffalo.edu/specialcollections/lovecanal/about/chronology.php

3. Though Love Canal residents engaged in both institutional public participation processes and alternative social protests, we limit our analysis to the indecorous tactics employed at public meetings and public hearings. This analysis is based on historical texts documenting the struggle at Love Canal. These include Levine's (1982) in-depth study of the unfolding process (which includes firsthand interviews with Gibbs and other LCHA activists), as well as Gibbs's (2011) memoir and handbook for environmental health organizing. Also, the documentary *The Poisoned Dream: The Love Canal Nightmare* provides real-time footage of many of the public hearings.

4. This analysis is based on transcripts received from the UDEQ's DAQ of received comments and field notes from two of the three authors who attended the event. This triangulation of data from in situ participatory research, public comments, and field notes—on top of swaths of research from newspapers and government records—provides a level of qualitative and rhetorical saturation that exceeds traditional textual analyses. For instance, actually attending this event allowed the authors to observe the micro-performances of participants through applause and stages of protest used during comments, such as signs and children; and to use facial

expressions, sounds, smells, and felt momentum as rhetorical data for our analysis. We defend that this rhetorical methodology is consistent with what Middleton, Senda-Cook, and Endres (2011) call Rhetorical Field Methods.

References

Bauman, J. (2007, January 11). Divine Strake session criticized. *Deseret News*. Retrieved from http://www.deseretnews.com/article/650222028/Divine-Strake-session-criticized.html?pg=all

Beierle, T. C., & Cayford, J. (2002). *Democracy in practice: Public participation in environmental decisions*. Washington, DC: Resources for the Future.

Bennion, K. (January 21, 2013). Utah cities atop EPA's worst-air-quality list. *The Salt Lake Tribune*. Retrieved from http://archive.sltrib.com/story.php?ref=/sltrib/news/55670441-78/utah-quality-logan-degrees.html.csp

Blum, E. D. (2008). *Love Canal revisited: Race, class, and gender in environmental activism*. Lawrence, KS: University Press of Kansas.

Bowers, J. W., Ochs, D. J., Jensen, R. J., & Schulz, D. P. (2009). *The rhetoric of agitation and control* (3rd ed.). Long Grove, IL: Waveland Press.

Bucchi, M., & Neresini, F. (2007). Science and public participation. In E. J. Hackett, O. Amsterdamska, M. E. Lynch, J. Wajcman, & W. E. Bijker (Eds.), *The handbook of science and technology studies* (3rd ed., pp. 449–472). Cambridge, MA: MIT Press.

Callister, D. C. (2013). Land community participation: A new "public" participation model. *Environmental Communication, 7*(4), 435–455. doi:10.1080/175240 32.2013.822408

Cicero (1961). *De Officiis* (W. Miller, trans.). Cambridge, MA: Harvard University Press.

Campbell, K. K. (1973). The rhetoric of women's liberation: An oxymoron. *Quarterly Journal of Speech, 59*(1), 74–86. doi:10.1080/00335637309383155

Cathcart, R. S. (1978). Movements: Confrontation as rhetorical form. *Southern Speech Communication Journal, 43*(3), 233–247. doi:10.1080/10417947809372383

Charland, M. (1987). Constitutive rhetoric: The case of the peuple québécois. *Quarterly Journal of Speech, 73*(2), 133–150. doi:10.1080/00335638709383799

Chávez, K. R. (2011). Counter-public enclaves and understanding the function of rhetoric in social movement coalition-building. *Communication Quarterly, 59*(1), 1–18. doi:10.1080/01463373.2010.541333

Cherwitz, R. A., & Zagacki, K. S. (1986). Consummatory versus justificatory crisis rhetoric. *Western Journal of Speech Communication, 50*(4), 307–324. doi:10.1080/10570318609374240

Cox, J. R. (1982). The die is cast: Topical and ontological dimensions of the locus of the irreparable. *Quarterly Journal of Speech, 68*(3), 227–239.

Cox, R. J. (1999). Reclaiming the "indecorous" voice: Public participation by low-income communities in environmental decision-making. In C. B. Short & D. Hardy-Short (Eds.), *Proceedings of the Fifth Biennial Conference on Communication and the Environment* (pp. 21–31). Flagstaff, AZ: Northern Arizona University School of Communication.

Cox, J. R. (2013). *Environmental communication and the public sphere* (3rd ed.). Thousand Oaks, CA: Sage Publications.

Daniels, S. E., & Walker, G. B. (2001). *Working through environmental conflict: The collaborative learning approach.* Westport, CT: Praeger.

Davies, S. R., & Selin, C. (2012). Energy futures: Five dilemmas of the practice of anticipatory governance. *Environmental Communication: A Journal of Nature and Culture, 6*(1), 119–136. doi:10.1080/17524032.2011.644632

De Certeau, M. (1984). *The practice of everyday life.* Berkeley, CA: University of California Press.

Delicath, J. W. (2004). Art and advocacy: Citizen participation through cultural activism. In S. P. Depoe, J. W. Delicath, & M.-F. A. Elsenbeer (Eds.), *Communication and public participation in environmental decision making* (pp. 255–266). Albany: State University of New York Press.

DeLuca, K. (1999). *Image politics: The new rhetoric of environmental activism.* New York: Guilford.

Depoe, S., & Delicath, J. W. (2004). Introduction. In S. P. Depoe, J. W. Delicath, & M. A. Elsenbeer (Eds.), *Communication and public participation in environmental decision making* (pp. 1–10). Albany: State University of New York Press.

Depoe, S. P., Delicath, J. W., & M. A. Elsenbeer (Eds.). (2004). *Communication and public participation in environmental decision making.* Albany: State University of New York Press.

Dickson, M. (2007, January 31). The people speak: When the public is not welcome. *Catalyst.* Retrieved from http://166-70-249-138.ip.xmission.com/component/k2/item/144-the-people-speak-when-the-public-is-not-welcome

Dietz, T., & Stern, P. C. (Eds.). (2008). *Public participation in environmental assessment and decision making.* Washington, DC: The National Academies Press.

Divine Strake: Hatch, Matheson criticize meetings on blast. (2007, January 12). *The Salt Lake Tribune.* Retrieved from http://archive.sltrib.com/story.php?ref=/ci_4998529

Enck-Wanzer, D. (2006). Trashing the system: Social movement, intersectional rhetoric, and collective agency in the Young Lords organization's garbage offensive. *Quarterly Journal of Speech, 92*(2), 174–201.

Endres, D. (2009). Science and public participation: An analysis of public scientific argument in the Yucca Mountain controversy. *Environmental Communication: A Journal of Nature and Culture, 3*(1), 49–75.

Endres, D. (2011). American Indian activism and audience: Rhetorical analysis of Leonard Peltier's response to denial of clemency. *Communication Reports, 24*(1), 1–11. doi:10.1080/08934215.2011.554624

Endres, D. (2012). Sacred land or national sacrifice zone: The role of values in the Yucca Mountain participation process. *Environmental Communication: A Journal of Nature and Culture, 6*(3), 328–345. doi:10.1080/17524032.2012.688060

Endres, D., & Senda-Cook, S. (2011). Location matters: The rhetoric of place in protest. *Quarterly Journal of Speech, 97*(3), 257–282. doi:10.1080/0033563 0.2011.585167

Endres, D., Senda-Cook, S., & Cozen, B. (2014). Not just a place to park your car: Park(ing) as spatial argument. *Argumentation & Advocacy, 50*(3), 121–140.

Fahys, J. (2007, January 11). Divine Strake visitors frustrated. *The Salt Lake Tribune.* Retrieved from http://archive.sltrib.com/story.php?ref=/news/ci_4990514

Fiorino, D. J. (1990). Citizen participation and environmental risk: A survey of institutional mechanisms. *Science, Technology, & Human Values, 15*(2), 226–243.

Fischer, F. (2000). *Citizens, experts, and the environment: The politics of local knowledge.* Durham, NC: Duke University Press Books.

Frosch, D. (2013, February 24). Seen as nature lovers' paradise, Utah struggles with air quality. *The New York Times.* Retrieved from http://www.nytimes.com/2013/02/24/us/utah-a-nature-lovers-haven-is-plagued-by-dirty-air.html?_r=0

Gibbs, L. M. (2011). *Love Canal and the birth of the environmental health movement.* Washington, DC: Island Press.

Governor sets Divine Strake hearings. (2007, January 10). *The Salt Lake Tribune.* Retrieved from http://archive.sltrib.com/story.php?ref=/news/ci_4984194

Gregg, R. B. (1971). The ego-function of the rhetoric of protest. *Philosophy & Rhetoric, 4*(2), 71–91.

Haiman, F. S. (1967). The rhetoric of the streets: Some legal and ethical considerations. *Quarterly Journal of Speech, 53*(2), 99–114.

Hamilton, J. D., & Wills-Toker, C. (2006). Reconceptualizing dialogue in environmental public participation. *Policy Studies Journal, 34*(4), 755–775. doi:10.1111/j.1541-0072.2006.00200.x

Hariman, R. (1992). Decorum, power, and the courtly style. *Quarterly Journal of Speech, 78,* 149–172.

Hendry, J. (2004). Decide, announce, defend: Turning the NEPA process into an advocacy tool rather than a decision-making tool. In S. P. Depoe, J. W. Delicath, & M.-F. A. Elsenbeer (Eds.), *Communication and public participation in environmental decision making* (pp. 99–112). Albany: State University of New York Press.

Hunt, K. (2014). "It's more than planting trees, it's planting ideas": Ecofeminist praxis in the Greenbelt Movement. *Southern Communication Journal, 79,* 235–249.

Hunt, K., & Paliewicz, N. P. (2013). "Are you *listening*?!" Indecorous voice as rhetorical strategy in environmental public participation. *Communication for the Commons:* 407.

Jamieson, D. (2007). The heart of environmentalism. In R. Sandler & P. C. Pezzullo (Eds.), *Environmental justice and environmentalism: The social justice challenge to the environmental movement* (pp. 85–101). Cambridge, MA: MIT Press.

Katz, S. B., & Miller, C. (1996). The low-level radioactive waste siting controversy in North Carolina: Toward a rhetorical model of risk communication. In C. Herndl & S. C. Brown (Eds.), *Green culture: Environmental rhetoric in contemporary America* (pp. 111–140). Madison, WI: University of Wisconsin Press.

Kinsella, W. J. (2004). Public expertise: A foundation for citizen participation in energy and environmental decisions. In S. P. Depoe, J. W. Delicath, & M.-F. A. Elsenbeer (Eds.), *Communication and public participation in environmental decision making* (pp. 83–98). Albany: State University of New York Press.

Klaus, M., & Mayhew, D. (2012, February 29). Kennecott causes one-third of air pollution. *The Salt Lake Tribune.* Retrieved from http://archive.sltrib.com/story.php?ref=/sltrib/opinion/53600999-82/rtk-pollution-utah-mining.html.csp

Lake, R. A. (1983). Enacting red power: The consummatory function in Native American protest rhetoric. *Quarterly Journal of Speech, 69*(2), 127–142. doi:10.1080/00335638309383642

Levine, A. G. (1982). *Love Canal: Science, politics, and people.* Lanham, MD: Lexington Books.

Matthews, K. W. (Director). (1997). *The poisoned dream: The Love Canal nightmare* [documentary]. New York: Films Media Group.

McGee, M. C. (1980). "Social movement": Phenomenon or meaning? *Central States Speech Journal, 31*(4), 233–244. doi:10.1080/10510978009368063

Middleton, M. K., Senda-Cook, S., & Endres, D. (2011). Articulating rhetorical field methods: Challenges and tensions. *Western Journal of Communication, 75*(4), 386–406.

Oravec, C. (1984). Conservationism vs. preservationism: The "public interest" in the Hetch Hetchy controversy. *Quarterly Journal of Speech, 70,* 444–458. doi:10.1080/00335638409383709

Penrod, E. (2016, Feb 22). Environmentalists frustrated as Utah air quality measures get stuck or cut. *The Salt Lake Tribune.* Retrieved from http://www.sltrib.com/home/3568453-155/environmentalists-frustrated-as-utah-air-quality

Pezzullo, P. C. (2007). *Toxic tourism: Rhetorics of pollution, travel, and environmental justice.* Tuscaloosa: University of Alabama Press.

Philips, L. J., Carvalho, A., & Doyle, J. (2012). *Citizen voices performing public participation in science and environment communication.* Bristol, UK; Chicago: Intellect.

Philipsen, G. (1992). *Speaking culturally: Explorations in social communication.* Albany: State University of New York Press.

Senecah, S. L. (2004). The Trinity of voice: The role of practical theory in planning and evaluating the effectiveness of environmental participatory processes. In S. P. Depoe, J. W. Delicath, & M.-F. A. Elsenbeer (Eds.), *Communication and public participation in environmental decision making* (pp. 13–33). Albany: State University of New York Press.

Stettler, J. (2011, May 5). Pollution fears cloud approval of Kennecott expansion. *The Salt Lake Tribune.* Retrieved from http://archive.sltrib.com/story.php?ref=/sltrib/politics/51752890-90/kennecott-expansion-mine-utah.html.csp

Treichel, J. (2001). *U.S. Department of Energy Public Hearing on the Possible Site Recommendation of Yucca Mountain.* Las Vegas, NV: U.S. Department of Energy.

Utah Department of Environmental Quality (DEQ). (2011a, February 22). *Summary—public hearing comments.* Available on CD-ROM. Retrieved January 8, 2016.

Utah Department of Environmental Quality (DEQ). (2011b). Kennecott Utah Copper LLC projects. Retrieved from http://www.deq.utah.gov/businesses/K/kennecott/cornerstone/index.htm#pubpart

Walker, G. B. (2007). Public participation as participatory communication in environmental policy decision-making: From concepts to structured conversations. *Environmental Communication: A Journal of Nature and Culture, 1*(1), 99–110. doi:10.1080/17524030701334342

Windt, Jr., T. O. (1972). The diatribe: Last resort for protest. *Quarterly Journal of Speech, 58*(1), 1–14. doi:10.1080/00335637209383096

Section II

Expanding Pathways
of Community Engagement

Chapter 7

Advancing Practical Theory in Environmental Communication

A Phronetic Analysis of Environmental Participation and Dialogue in New Zealand

GILES DODSON AND ANNA PALLISER

Introduction

International agreements such as the Convention on Biological Diversity (1992) and the Aarhus Convention (1998) highlight the importance of involving local people and using participatory approaches to environmental decision making. Participatory initiatives now are found around the world, including in New Zealand (Armitage, 2005; McCallum, Hughey, & Rixecker, 2007). Participation commonly involves local citizens working alongside representatives from government agencies and scientific institutions. These participants build trust, share understanding, and learn experientially as they work together to manage local socio-ecological systems (Berkes, Colding, & Folke, 2003; Plummer & Armitage, 2010). Such approaches may be considered to reflect the "dialogic turn" in environmental governance, "in which experts and target groups are reconfigured as participants in sites of dialogue where knowledge is co-produced through mutual learning" (Phillips, Carvalho, & Doyle, 2012, p. xx).

Ultimately, the success of environmental initiatives is dependent on people—the ways they work together, their values, and their aims (Cundill, Cumming, Biggs, & Fabricius, 2012). Yet in practice, participatory approaches may be only partially realized, with citizens having a limited degree of input (Phillips et al., 2012). Without considering sociopolitical-legal contexts, participatory approaches risk being considered panaceas (Conley & Moote, 2003). For example, valuing citizen knowledge and perspectives less highly than those of government agencies and scientists can undermine building collaborative relationships (Ulrich & Reynolds, 2010). Such inequities may result from unexamined assumptions. Healy (2009) discusses how scientists assume their role is to present facts, with citizen input limited to preferences and values; consequently, citizen knowledge is marginalized. Senecah (2004) finds government agency representatives believe citizens lack capability for governance and cannot understand the complex issues involved, which is likely to lead to citizen input being devalued. Additionally, citizen power to effect local change may be limited by national legislation, which excludes citizens from final decision making (Dodson, 2014b).

Deliberative processes, where participants freely discuss assumptions about how different knowledge, values, and perspectives are used, are posited by some to be a requisite for effective collaboration (Popa, Guillermin, & Dedeurwaerdere, 2015; Ulrich & Reynolds, 2010). As Reed (2008, p. 2417) argues: "stakeholder participation needs to be underpinned by a philosophy that emphasizes empowerment, equity, trust and learning." By engaging with these issues, research can play an important role in participant deliberations, providing information and insights that may help build collaboration.

A methodological approach that focuses research onto these issues is phronetic social science (Flyvbjerg, 1998; 2001). Phronetic social science focuses analysis on the issues that matter to people involved in management situations. It is primarily concerned with offering insights that may be of value to practitioners rather than developing generally applicable or predictive theory. In this chapter, the phronetic approach engages with the ways different knowledge, values, perspectives, and power relationships play out in two specific cases of environmental management. A Foucauldian/Nietzschean approach to power is followed, as discussed in Flyvbjerg (2004, p. 293).[1] Thus, power relationships are considered to be inherent in all social relations, best dealt with by making them transparent so that involved participants can then address them.

In this chapter, we demonstrate the value of a phronetic method in both developing and analyzing public participation. We argue that questions of power cannot be left out of participation. Practical concerns for how power structures participation provide practitioners with tools to more deeply appreciate the complexities of participation and to design and implement democratic and sustainable participatory projects. On the one hand, the phronetic approach provides a practical guide, but on the other hand, it enables a critical engagement with power as it shapes that context in which participation takes place. One of our case studies focuses on a marine reserve initiative and the other on an integrated catchment management project. Both contain important participation of tangata whenua[2] and the inclusion of mātauranga Māori (Māori knowledge) in environmental management. The chapter initially discusses our theoretical and methodological approaches and then outlines the social, cultural and policy contexts in which our case studies are embedded. Our case studies and their analyses are presented, illustrating the value of the phronetic approach, particularly by drawing attention to the power relationships inherent in participatory process. These cases highlight that high-quality participation and the potential environmental gains thereof may be jeopardized unless power relationships are adequately recognized and addressed.

Participation and Dialogue

Walker (2007) and others (Phillips, 2011; Senecah, 2004) have criticized the ways "participation" has become institutionalized into established processes or sets of principles. For example, formal consultation processes, where citizens are invited to speak at public hearings and to write submissions, may be ineffective, leaving citizens feeling unheard (Senecah, 2004). Public participation may be structured in ways that do not promote democratic outcomes (Phillips et al., 2012; Horsbol & Larssen, 2012), and contested outcomes are in the nature of deliberative democracy (Hagendijk & Irwin, 2006). Thus, environmental communication should be concerned with the practical experience and outcomes of participation, rather than idealizing process or concepts (Cox, 2007). If participatory processes are idealized and imposed on citizens from above, environmental initiatives may be undermined by creating resistance and disempowerment at the local level (Adger, Brown, & Tompkins, 2005). This may be of concern when participants have different environmental priorities and values, as is frequently the case in postcolonial societies such as New Zealand.

Some scholars contend that building theory about participatory processes is difficult, given the range of contexts, values, perspectives, and power disparities embedded in participatory situations (Anderies, Walker, & Kinzig, 2006). Senecah (2004, p. 21) considers that theoretical approaches need to be "useful and flexible in practice." Consequently, "practical" theory is required to understand and analyze public participation, which should be focused on building trust among participants rather than on building general theory. Setting aside theory building (Senecah, 2004), the issue we are left with therefore is how values, voices, and relationships are reconciled in participation, made meaningful through dialogue.

Dialogue implies trust building via respect for and inclusion of the values of participants. It also implies recognition of the processes of meaning-making that underpin the values held by different cultures and participating groups (Dutta, 2011, pp. 37–38). As Servaes (2008, p. 96) suggests, dialogic communication forms the normative basis of participatory social change. Dialogue is therefore seen as the process through which participatory relations are constituted and democratic outcomes enabled (Singhal, 2001).

Likewise, Phillips (2011) proposes that the discourse of "dialogue and participation" connotes equitable, democratic relations in which dialogue and action are directed toward social equity (p. 59). For Phillips (2011), dialogue possesses a relational quality, where dialogue is understood as "horizontal processes of information exchange and interaction" (Morris, in Phillips, 2011, p. 65). More hierarchical processes of participation, such as the formal public consultation processes, may be promoted as open and participatory yet in practice be controlled by an elite group holding decision-making power. Brulle (2010) argues that environmental *communication* should be directed toward supporting civic engagement and "scientific citizenship," in which citizens become involved in science and ecological and policy processes (including decision making), particularly through articulating alternative discourses of community knowledge and values in relation to the environment.

Critical engagement with participatory processes is essential whether these processes are hierarchical in nature or whether they aim for a more equitable and democratic approach. Substantive questions remain regarding the nature of public participation in environmental governance. Different circumstances and contexts as well as localized configurations of power, interests, and resources can confound what otherwise may be effective dialogic process and sound policy prescriptions. Senecah (2004) suggests participation processes should aim for the "trinity of voice" so that all stakeholders have access, standing, and influence. Our question is, "Do they have this in practice?"

Practical Theory: Phronetic Social Science

Now that citizens have been reconfigured as participants in environmental decision making alongside experts and governance elites (Philips et al., 2012), participant deliberations about how different knowledges and perspectives should be valued become imperative (Ulrich & Reynolds, 2010). Consequently, the role played by research (which is traditionally seen as providing elite knowledge) requires questioning. Flyvbjerg (2001) argues for reconfiguring the role of social science, focusing on phronesis, the practical and situated wisdom of people working on the ground.[3] Here phronesis, or practical, values-based rationality, is contrasted with epistemic (scientific) and technical rationalities frequently deployed by governance and administrative elites. An administrative or managerial preference for elite-produced frameworks, models, or processes for how things should be done often marginalizes phronetic knowledge gained from practical experience. Flyvbjerg (2001, 2005) and others (Schram & Caterino, 2006) argue that methodological approaches that assume research produces theory (applied in top-down fashion to practice) are less valuable than approaches aiming to "produce food for thought for the on-going process of public deliberation, participation and decision-making" (Flyvbjerg, 2005, p. 39). Research outputs, such as this one, are therefore aimed at groups of participants and should encourage deliberating the value of the research in relation to their specific circumstances. As Flyvbjerg (2001, p. 139) says, "phronetic research is dialogical in the sense that it includes, and, if successful, is itself included in, a polyphony of voices, with no one voice, including that of the researcher, claiming final authority."

For Flyvbjerg (2001), phronetic research should focus on the practical issues concerning people who are working together in particular localities at particular times. While phronetic social science does not so far appear to have engaged with environmental management or communication studies, we argue it has value as a practical theory deployed to bridge the spaces between theory and practice. Its purpose is to clarify "where we are, where we want to go, and what is desirable according to diverse sets of values and interests" (Flyvbjerg n.d., para 6). As method and practical philosophy (Flyvberg, 2001, pp. 55–60), phronesis is concerned with the ethics of action; it is a practical, values-based rationality guiding social action. Thus, it can help to describe and evaluate what is happening in practice in a given context and provide practical guidance to participants.

This approach answers the questions "Where are we going?" "Who gains and who loses, through what kinds of power relations?" "Is this desirable?" "What should be done?" (Flyvberg, 2001). Through examining lived experience of participants, this approach has the potential to be both emancipatory—in line with the normative tenets of environmental communication as a "crisis discipline" (Cox, 2007)—and dialogic/deliberative through the inclusion of diverse interests. In asking "What should be done," attention is focused on the importance of deliberating diverse values when dealing with complex issues, in which all values and knowledges involved tend to be perspective dependent, incomplete, and contested, including technical rationality (Funtowicz & Ravetz, 2003; Ulrich & Reynolds, 2010). In analyzing participatory environmental communication, attention is drawn to practical questions of how voices[4] are enabled within deliberative process and with what effect. In other words, how does this combination of voices shape knowledge of the environment and decision making?

A phronetic approach also draws attention to the power relationships inevitably embedded in both participatory processes and the broader policy context in which these ultimately reside (Flyvbjerg, 2001). Thus, this method permits analysis of the practical outcomes of participation, moving beyond normative conceptions of deliberative processes and locating the values-rational, contextual, and diverse production of environmental knowledge within the context of ever-present relations of power (Flyvbjerg, 2001, p. 117). In doing so, phronesis elevates values-rationality, nontechnical, and indigenous knowledge to the same level as technical rationality.

Participatory Environmental Management and Mātauranga Māori in Aotearoa New Zealand

High-profile examples exist in New Zealand of participatory groups wielding significant authority and influence in resource management, particularly in areas of fraught and competitive resource allocation, such as freshwater (Canterbury Water Management Strategy, 2009;[5] Land and Water Forum, 2010; 2012a; 2012b; Waikato River Authority, 2013) and marine protected areas (Department of Conservation, 2005). These initiatives show emergent institutionalized participation; however, they typically sit outside the mainstream environmental management framework.[6] While some approaches provide for the inclusion of community, Māori, and other stakeholder perspectives (RMA, 1991[7]), most are consultative rather than collaborative or participatory. Where collaborative approaches are being pioneered, as in the management of Canterbury freshwater, the legislative framework underpinning these is uncertain.

Mātauranga Maori is increasingly recognized as an important component in environmental management and policy (Moller et al., 2009; Mutu, 2010). Mātauranga Māori is concerned with Māori worldviews, values, systems of knowledge, and spirituality. It provides an integrated framework for addressing the responsibilities and relationships inherent in contemporary decision making. Mātauranga Māori can be understood as related to traditional ecological knowledge (TEK) (Moller, 2009). As Berkes (2008) describes, TEK is the body of knowledge, practice, and belief, handed down through generations by cultural transmission, concerning the environment and the human place within it. Mātauranga Māori, however, should not be mistaken for ossified, traditional knowledge, as it is creative, flexible, and adaptive, applicable to all areas of human activity and knowledge (Harmsworth & Awatere, 2013).

Mātauranga Māori is especially applicable with respect to environmental management (Harmsworth & Awatere, 2013; Kawharu, 2000; Marsden, 1992; Rotorangi and Russell, 2009). Key environmental concepts cover not only the use or protection of the natural world, but also the genealogical, cosmological, and philosophical underpinnings of resource management (Durie, 1998; Kawharu, 2000; Harmsworth & Awatere, 2013). For example, connect the human, natural, and spiritual realms by tracing whakapapa (loosely, genealogy); thus, Māori are related, with the rights and responsibilities implicated in blood relationships, with the whole web of life (Harmsworth & Awatere, 2013). Clearly there are tensions between western scientific models of supposedly values-free empirical and hypothesis focused methods, and qualitative, context-specific, spiritually imbued and values-driven mātauranga Māori. However some see mātauranga Māori to be dynamic and evolving knowledge, complementary to western scientific knowledge (Harmsworth and Awatere, 2013; Mutu, 2010; Tipa, 2010).

Productive partnerships in which mātauranga Māori and Western science address pressing environmental issues are apparent across the gamut of socioenvironmental contexts. These include the rights and responsibilities of environmental guardianship, or kaitiakitanga (Kawharu, 2000; Marsden, 1992; Mutu, 2010); to assessments of traditional food resources (Moller et al., 2009); wetlands (Forster, 2010); the protection of Māori heritage assets (Kawharu, 2000); and river health (Tipa, 2010).

Increasingly Māori are insisting on their rights to involvement in environmental management (Tipa & Welsh, 2006; Tūhoe Deed of Settlement,

2012; Waitangi Tribunal, 2011). The Resource Management Act (RMA) of 1991 requires the inclusion of Māori perspectives in environmental decision making, and recent Waitangi Tribunal (2011) findings recommend deepening Māori environmental governance roles.[8] However, ongoing grievances and mistrust resulting from historic and contemporary injustice looms large in all considerations of postcolonial approaches to environmental policy in New Zealand. A significant and growing body of Māori research critiques participatory processes in environmental governance from the perspective of Māori rights (Coombes and Hill, 2005; Kawharu, 2000; Mutu, 2010). Consequently, it is also necessary to critically interrogate moves toward "partnership" or collaborative modes of environmental management that seek to incorporate mātauranga Māori or to engage Māori as partners to avoid "partnership" and "participation" becoming idealized narratives presenting a panacea for multiple and persistent grievances (Conley & Moote, 2003).

Case Studies: Multi-Stakeholder Partnership and Marine Reserve Campaigning

The following two case studies exemplify many of the issues involved with current participatory approaches to environmental management in New Zealand. The first study, of an integrated approach to harbor/catchment management, is a positive example of how diverse participants in environmental governance are engaging with a large-scale collaborative initiative. A polyphony of voices are united by this initiative, which include mātauranga Māori, western scientific knowledge, local communities, and government agency representatives, in equitable ways. Thus, the practical wisdom, or phronesis, of tangata whenua (Indigenous Māori) and of non-Māori locals (who may also have intimate connections with local natural resources) is respected and given a voice in the various arenas of this initiative.

In contrast, the second case study concerns a campaign to establish a coastal marine reserve and is focused on barriers to effective collaborative initiatives. Like the first study, this case illustrates the determination of government agency representatives and tangata whenua to include Māori and western perspectives equitably in governance arrangements, and the phronesis of tangata whenua played an important part in the project. However, despite this, the outcomes were not positive; inequitable power relationships ultimately undermined this initiative.

Case Study One: Integrated Kaipara Harbour Management

The Integrated Kaipara Harbour Management Group (IKHMG) is a multi-stakeholder partnership (Warner, 2007) established in 2005 with the central aim of promoting interagency coordination and management and the use of mātauranga Māori in restoring Kaipara Harbour and its catchment. Kaipara Harbour is New Zealand's largest harbor. Its catchment, which encompasses 640,000 ha, extends from the northwestern reaches of Auckland city to the north of Whangārei. Kaipara Harbour is the central taonga (treasured possession) of Ngāti Whatua, the confederation of Māori hapū (subtribes) who maintain mana whenua and mana moana (traditional authority over land and sea) over this region. The IKHMG is led by Ngāti Whātua and draws together stakeholders from governance, policy, industrial, community, and nongovernmental sectors. It has articulated a guiding, common vision to create a healthy and productive Kaipara Harbour (IKHMG, 2011).

Central to the epistemological foundation, practice, and operation of the IKHMG is the integration of Western sciences with mātauranga Māori. As a collaborative, multi-stakeholder partnership, the key aims of the group are to gather knowledge, to develop innovative approaches to catchment management, to support community action, and to influence policy and regulation affecting Kaipara Harbour, across its priority areas.[9]

Since 2005, the IKHMG has successfully built the "stakeholder partnership" model, using the impetus gained from recent treaty settlements[10] and associated negotiations with the Crown to build engagement with regional stakeholders. Central to "integrated management" is bringing together disparate and disconnected regulatory tools to manage the harbor and catchment as a social-ecological system (IKHMG, 2011). The engagement and participation of these stakeholders have been a notable success of the IKHMG. Importantly, since 2012, the Kaipara Harbour Joint Political Committee has been established, bringing together the region's territorial authorities (councils) to integrate harbor and catchment management.[11]

IKHMG is a high-level stakeholder partnership, engaging directly with governance, regulatory, and scientific partners, yet it is also concerned with encouraging community participation. In 2014, an extended campaign raised community awareness about the pressures on the harbor, informing the public about the IKHMG and encouraging participation at a community symposium. This event was a demonstration of community support for the IKHMG and its agenda for harbor and catchment management and combination of western, science-based approaches with mātauranga Māori.

The two-day symposium followed a conference format featuring approximately 40 separate presentations under several themes:

- ecosystems;

- connecting with mātauranga Māori;

- integrated management;

- communities and relationships;

- implementing integrated management and planting two million trees (a central IKHMG project).

Community members, practitioners, landowners, and indigenous speakers delivered symposium presentations and interactive sessions alongside technical experts and officials. The event provided opportunity for discussion of issues facing the harbor, barriers to effective management, successes achieved in realizing the IKHMG vision, and future work. The multiple voices constituting the forum were constitutive of this dialogic space. During a final session, participants created "idea-walls" to capture participant contributions.[12] These contributions were recorded and incorporated into IKHMG strategic planning. However, how this collaboration will be translated into practical environmental management and governance, given the absence of formal recognition of collaborative management within the existing legislative and regulatory framework, remains to be seen.

CASE STUDY TWO: PARTNERSHIP AND MARINE PROTECTION
AT MIMIWHANGATA

Mimiwhangata, located on the northeast coast of Northland, North Island, New Zealand, is an area of ecological, scientific, and cultural importance to Māori and non-Māori. The history of terrestrial and marine conservation dates from the 1970s, when ecological surveys were first conducted. These intensively studied this area's outstanding natural and ecological features (Ballantine et al., 1973; Grace, 1981). The Northland east coast is both an ecologically high-value marine environment and a region of high-intensity recreational and commercial use. In the early 2000s, Mimiwhangata was identified as a focus for marine protection by the New Zealand Department of Conservation (DOC). From the DOC's perspective, a goal was to "get it right" in relation to collaboration with local Māori. If a marine reserve could be established at Mimiwhangata based on collaborative management/ governance principles and founded on a constructive partnership, then this could potentially provide both a model and precedent for future marine reserve establishment (Dodson, 2014).

Marine reserves, which are "no-take" areas, comprise the highest level of marine protection in New Zealand. However, the present legislative regime makes no reference to participatory, collaborative, or indigenous concepts. The Marine Reserves Act of 1971 provides for the preservation of unique marine environments for scientific purposes, thus undermining the exercise of kaitiakitanga (customary environmental relationships involving sustainable use), which is central to both customary practice and self-determination (rangatiratanga) (Kawharu, 2000). In addition, this act provides for little in the way of devolved or collaborative governance over marine protected areas (Uunila, 2003). The Conservation Act of 1987 does provide for "advisory committees" to advise the minister of conservation in relation to certain conservation areas, including marine reserves. However, these bodies have little in the way of real governance authority, and committee members elsewhere have voiced resentment at their lack of decision-making power (Uunila, 2003).

In the process of early consultation (beginning in 2001), a leadership group centered on local hapū (sub-tribe) Te Uri o Hikihiki quickly emerged, composed of local elders who were concerned both for the local marine environment and wider socioeconomic issues facing the area. Commencing in 2002, DOC and Te Uri o Hikihiki undertook a lengthy engagement process.[13] The project partners held an ongoing series of "working group" meetings and other hui (meetings) continuing until 2006. In these discussions, the need for some form of marine protection at Mimiwhangata was quickly agreed on. The form that protection should take and the governance of that institution—a central concern for Māori—quickly emerged as the focus for tangata whenua as they sought the meaningful restoration of their authority over the area and involvement in its management.

As part of the dialogue, a senior kaumatua (elder) and local leader made a public statement at Mimiwhangata. He expressed his concern over the degradation of the local marine environment and the depletion of fish stocks, calling for a rāhui (temporary closure)—a tikanga Māori (Māori protocols and customs) form of temporary closure—over the Mimiwhangata area for a period of 25 years (DOC, 2004). This declaration of a rāhui by a senior kaumatua was considered of fundamental importance both to the

DOC, as an expression of support for its policy, and for local people, for whom purposeful traditional leadership was evident. Although rāhui possess only very limited statutory status,[14] the public enunciation of this measure carried significant customary and cultural importance. This public declaration permitted legally sanctioned marine protection measures to be meaningfully endorsed in culturally appropriate terms (Dodson, 2014).

Tangata whenua envisioned a governance structure in which ultimate authority and decision-making responsibility rested with them, particularly given the restrictions on customary fishing that the reserve would require. While an advisory role, such as that provided by section 56 of the Conservation Act of 1987, may have provided some degree of involvement in reserve management, this fell short of local expectations that Māori authority be fully recognized. Thus, it presented a significant barrier to collaborative, adaptive marine protection using shared values and integrated knowledge systems.

It is important to note also the delicate balance that existing governance frameworks required project partners to maintain. As the then DOC area manager emphasized, the department's pragmatic approach to establishing joint governance rested on achieving a workable arrangement within existing frameworks and using that as a foundation on which to build support for more progressive forms of reserve governance over time. Consequently, although officially "advisory committees" possess limited authority, in practice these bodies could become important trust-building institutions and vehicles for tangata whenua involvement (Dodson, 2014).

Ultimately, it was decided by consensus among project partners that Te Uri o Hikihiki would be a joint applicant with the DOC in the formal application process. The question of governance remained unresolved. Nonetheless, the tangata whenua partner identified being a joint applicant as a firm opportunity to advance its strategy for hapu empowerment and development, with a clear vision of restored kaitiakitanga (customary relationships) and enhanced rangatiratanga (self-determination) at the top of its agenda (Dodson, 2014). Furthermore, both parties recognized that if traditional relationships and authority were restored through innovative governance frameworks, a powerful sense of local empowerment would be achieved while also delivering marine protection outcomes.

Despite these efforts and the establishment of a strong, focused partnership, ultimately the marine reserve campaign has stalled. On the one hand, unanimous community support was not achieved among tangata whenua, many of whom maintained reservations over ultimate reserve governance structures. On the other hand, and more crucial to the reserve application, was the 2006 decision to place the application to establish a marine reserve at Mimiwhangata "on hold" because of the promulgation of a broader Marine Protected Areas (MPA) policy by the New Zealand government. This was a political decision made at a senior level within the DOC. The government's MPA policy has provided comprehensive direction for the national marine environment and an integrated, regional forum-based approach to marine reserve establishment. While its intent has been to institutionalize collaborative and adaptive management approaches toward marine protected areas, this policy has to a large extent halted the establishment of government agency–sponsored marine reserves on high-use/high-value coastlines, such as the Northland east coast.

With What Effect? Communicative Spaces in Practice

Flyvbjerg (2005, p. 40) suggests the following four questions to contribute a phronetic approach to research. The questions focus analytical attention on the practicalities, process, and power relations that characterize participatory environmental management. These questions are (1) Where are we going?; (2) Who gains and who loses, and through what kinds of power relations?; (3) Is this development desirable?; and (4) What, if anything, should we do about it?

The term "we" in the above questions refers to anyone who has interest in or involvement with environmental management deliberations in the location (including researchers). Question 1 leads to probing the past as well as the future to examine the trajectory of our direction and demands critical analysis of decision making and its contexts. Question 2 probes power relationships, which may result in certain groups or knowledges being privileged or marginalized (these may be perceived in different ways according to different perspectives). Examining the past, where we are heading, and any power disparities provides the opportunity to question dominant patterns of governance and decision making (Hargreaves, 2012). Question 3 leads to asking "desirable for whom?" and as such encourages exploration of the different perspectives and values operating within the participatory group (including the perspectives of researchers), as well as exploring if the development is desirable for the local environment. The final question addresses recommendations to be made, which may arise from research and/

or from the participatory group. Research findings are thus viewed as one perspective among many, available to be further deliberated regarding their value by the communities involved.

WHERE ARE WE GOING?

Because of the IKHMG's efforts, the Kaipara Harbour community appears to be heading along a trajectory of increased and effective multi-stakeholder participation in resource management, with mātauranga Māori approaches and tools prominent in this work. In this case study, community-driven communication strategy and symposia are clearly shown to be effective tools for fostering community involvement and civic engagement in deliberating resource management issues. While elements of "campaign communications" were present, ultimately these too were community developed and delivered, and in support of the grassroots institution of the community symposium. Here the community constructed its own dialogic space, in which active citizenship could be demonstrated and communicated outside elite design or control (Brulle, 2010).

Such processes institute "horizontal" (Phillips, 2011) dialogue, in which community members (both indigenous and nonindigenous), governors, representatives of industry, regulators, and scientists are engaged in equitable relations. As with the IKHMG symposium, horizontal dialogue can produce the conditions for the legitimate inclusion of "value-rationality" in deliberations—Flyvbjerg's (2001) polyphony of voices. Under these conditions, it may be possible to approach requirements that Senecah (2004) articulates for effective participatory process: that relations of trust are established, that participants are legitimized, and that there is the possibility of their voices being recognized as important and influential. The community symposium did provide these conditions, and community members and symposium participants were legitimate voices in deliberations concerning the harbor and its management. Indeed, the IKHMG event supported the efflorescence of value-rationality (Flyvbjerg, 2001), as community and indigenous voices were included, legitimated, and valued in these discussions.

At Mimiwhangata, the future trajectory for community participation in marine conservation appears unclear. The marine reserve campaign has been put on hold. The long deliberative process undertaken by tangata whenua and the DOC has resulted in no concrete outcomes, which has been disheartening for the many people who invested significant time and energy into the project. The long-established norm of marine reserves being managed solely by the DOC was challenged by this initiative, in which both the DOC and tangata whenua sought to establish more participatory reserve management approaches, to establish trust, and to advance innovative marine protection strategy. However, ultimately this challenge was unsuccessful, and while the work done during this initiative may feed into future initiatives, it is equally likely that the failure to establish the reserve at Mimiwhangata will contribute to tangata whenua viewing further joint initiatives with caution,[15] which may be a serious setback for marine conservation in New Zealand.

A key reason given for the reserve campaign being "put on hold" is the promulgation by the government of the MPA policy, which seeks to take a regional, integrated, and forum-based approach to what historically has been a highly contentious process of campaign-driven reserve establishment. This political move has meant the stalling of marine protection measures being established on the North Island east coast, a high-value, high-use coastline. A successful MPA approach has been put into effect on the South Island West Coast (Ministry for Primary Industries, 2015); however, this coastline is much less intensely accessed and used by the public, fishers, and Māori. How the MPA process will "roll out" in Northland remains uncertain and politically sensitive (given the intensity of public use and competing values with respect to the marine environment). In the meantime, significant goodwill established through the ongoing efforts of local Māori and of the DOC to work together has been seriously undermined, and the advance of marine protection on this high-value coast is not proceeding.

Who Gains and Who Loses, and by Which Mechanisms of Power?

The decision by the DOC to put the Mimiwhangata marine reserve on hold resulted in significant losses for the community and local DOC representatives. Tangata whenua leaders had worked long and hard to achieve a cautious agreement from their communities to go ahead with a reserve application, despite no formal recognition of tangata whenua governance power in reserve management. With no positive outcome resulting from this work, it could easily be considered as another example of government agencies overriding tangata whenua attempts to play a more active role in managing natural resources. Notwithstanding the provision for "advisory" input from tangata whenua, clearly the currently available marine reserve governance structures position tangata whenua as less powerful than the DOC. The declaration of a rāhui, in accordance with traditional and

customary authority, also highlights the power disparities between government agencies and tangata whenua. Rāhui and marine reserves have similar
protective approaches; however, the limited statutory status of rāhui (and
not the type of protection it is designed to offer) renders mātauranga Māori
tools inferior to conventional forms of environmental protection, such as
reserves and conservation legislation. A disregard for Māori values is contained within these governance tools. The Marine Reserves Act of 1971,
for instance, fails to provide for Māori relationships and practices geared
toward sustainable use. As noted, the current marine reserves legislation
provides "no-take" protection for explicitly scientific purposes. Clearly, this
power relationship marginalizes Māori marine governance tools over those
designed by government agencies.

The national-level policy change was a crucial factor in putting the
reserve on hold. The way centralized policy changes take precedence over
long-term, local-level decision making requires scrutiny. This is especially
the case with regard to Māori, whose governance practices are community
based, deliberative, consensus focused, and centered on specific areas where
traditional authority can be exercised. This central policy change also resulted
in local DOC staff losing, not only in terms of the work they had put
in to develop the marine reserve proposal, but also in undermining their
collaborative relationships with tangata whenua. In addition, local ecosystems lost the opportunity for increased protection, and local communities
lost the opportunities for the increased tourism that is often the result of
marine reserve establishment (Cocklin, Craw, & Mcauley, 1998), as well as
related spillover effects from the marine reserve establishment (Russ, Alcala,
Maypa, Calumpong, & White, 2003). Overall, a long-term strategy for
partnership-based marine protection involving Māori and DOC in Northland has been jeopardized.

The "Looking back . . . thinking forward" community event of the
IKHMG provides a useful counterpoint to the agency-led marine reserve
campaign at Mimiwhangata and provides a possible model for the communication of the complementarity of indigenous knowledge—in our case,
mātauranga Māori—and western science. The purpose of the IKHMG
and its community engagement has been to develop and realize a "common vision": a healthy and productive Kaipara Harbour (IKHMG, 2011)
and an articulation of community and stakeholder values and narratives
in relation to the harbor. Dialogic and participative engagement serve as
vital mechanisms through which values-based rationalities and narratives

are produced, articulated, and recognized. Indeed, in this way, otherwise frequently marginalized voices (Dutta, 2011) are given equal space, time, and consideration alongside established or institutional voices, such as governance authorities or scientists. This is partly due to leadership by Māori, thus reversing the normal agency-led development of similar initiatives in New Zealand. Consequently, the mana (authority, influence, or status) of indigenous voices is affirmed, and the indigenous conceptual foundation for engaging in participatory and deliberative environmental communication is included. In this case, the presence of indigenous leadership is central and vital to the formation of the IKHMG and the bringing together of the community at events such as the symposium.

The IKHMG is clearly facilitating equitable sharing of differing knowledge and values. It nonetheless remains to be seen how these will be translated into practical decisions about harbor management. There is clear potential for governance decisions to be made that represent the full range of voices; however, this potential was also present in the Mimiwhangata deliberations. While collaboration remains unacknowledged within formal statutory arrangements in New Zealand, the risk remains that the motivation and empowerment engendered by the IKHMG will founder if stakeholders find their perspectives and knowledge devalued.

A key factor that differentiates the case studies is the institutional context in which each is situated and the consequences this can have for project control, development, and, ultimately, local empowerment. On one hand, the marine reserve campaigners, while seeking to maintain and restore traditional authority, were located within the bureaucracy, as the campaign was driven by the government's conservation agency, the DOC. On the other hand, although governance and regulatory agencies are IKHMG partners and contributors, the IKHMG is led by and takes its authority from the community, in particular Māori leadership; regulators and governance authorities are *participants*, positioned alongside others. The Mimiwhangata campaigners were able to access state resources and scientific expertise to advance their work; however, ultimately they were also vulnerable to capricious government policy. With the IKHMG project, leaders do not have access to either significant resources or political power. Remaining outside bureaucratic structures has meant that project control resides with the community leadership and is not at risk from changes to policy. Yet project leaders can maintain the participation of regulators and governors. Both cases present examples of differing reconfigurations of "decision-space" (with differing levels of success). In both cases, however, "decision-authority" ultimately continues to reside with statutory authorities, notwithstanding the attempts by participants to reshape this relation (Walker, Senecah, & Daniels, 2006).

IS THIS DEVELOPMENT DESIRABLE?

The phronetic social science method encourages both researchers and practitioners to explore the value of what is happening in practice. Consequently, as researchers, we examine the desirability of the trajectory of each case. We do so in the context of what we know about participatory approaches to environmental management and from the perspectives of involved participants. Certainly the bottom-up development of inclusive and equitable participatory processes resulting from the IKHMG and the "Looking back. . . . thinking forward" symposium appears desirable, especially in a country where there have been significant conflict and criticism about the marginalization of tangata whenua in resource management (Waitangi Tribunal, 2011). By contrast, however, developments at Mimiwhangata appear far less desirable. The Mimiwhangata case points to the pitfalls and weaknesses of participatory approaches embedded in unsupportive governance or legislative contexts. Looking at both a grassroots level and an administrative/regulatory level, these cases illustrate the desirability of participative, dialogic approaches, especially under conditions of diversity and complexity. Both cases also illustrate the viability of driving environmental communication and management according to community-based values-rationality. And, although they have not directly informed each other, participants in both cases are aware of the difficulties of achieving their aims in unsympathetic regulatory and governance contexts.

Partly because of IKHMG efforts, a joint political committee has been established since 2012 by Kaipara Harbour governance authorities (Auckland Council, Northland Regional Council, and Kaipara District Council) along with Māori representatives to advise on governance issues relating to the harbor. This illustrates local government and iwi (tribal) recognition of the need for integrated approaches to the harbor and catchment management. However, this political committee has no mandate, decision making, or governance authority. The existing statutory framework for resource management in New Zealand does not constructively provide for the possibility of integrated management, which is clearly undesirable in terms of this case study. Collaborative or adaptive management strategies are only weakly provided for under the Resource Management Act of 1991, and the most advanced attempt to manage resources according to community-level, collaborative, and values-driven approaches, the Canterbury Water Management Strategy, exists within an uncertain statutory framework (CWMS, 2009, pp. 56–59).

The Kaipara district is divided between four territorial authorities and multiple sectoral regulators, with at best an emergent framework to provide for collaborative and integrated management. This increases the risk that despite equitable sharing of knowledge, values, and perspectives via the IKHMG, actual management decisions, which must comply with protocol established under resource management legislation, may privilege certain knowledges and perspectives over others. In the absence of enabling legislation, future management decisions will inevitably be made by local government under the Resource Management Act of 1991, despite high-quality and progressive participatory engagement of the community. This is clearly undesirable for stakeholders who have invested in the IKHMG process in the expectation of direct involvement in decision-making processes. This may especially be the case for Māori, who have engaged previously with collaborative initiatives between the state and tangata whenua in the expectation of full involvement in management, only to find themselves excluded (Tipa & Welch, 2006).

Likewise, the Mimiwhangata study illustrates the undesirability of marine conservation legislation that does not adequately provide for inclusive participatory processes. In addition, from a local perspective, the new MPA policy is undesirable both for local DOC agents and for tangata whenua, as it may undermine hard-won relationships and understanding. However, from the perspective of higher levels of DOC management structures, this development may be desirable because by putting the reserve on hold, there is the opportunity to ensure that reserve establishment in the region will comply with what is envisioned under the MPA policy. While this is a "neater" solution, given the phronetic concern for the contextual, local, and specific, a top-down approach to marine protection runs the risk of applying a centrally designed and promoted model to local conditions, which may be both inappropriate and undermining of local values. This analysis points to the need for a diversity of flexible approaches for marine protection, designed to be suitable for particular places at particular times (Ostrom, Janssen, & Anderies, 2007). In the meantime, tangata whenua remain excluded from environmental decision making, and the local marine environment does not enjoy high-level protection.

Conclusion

Having drawn attention to the issues associated with these two cases, "What should be done?" becomes an imperative question. Our analysis of these two cases illustrates that the community-based events and communication processes used by participants have been constructive and should continue to be supported. Notwithstanding the limitations of the approaches presented here, clear lessons for both community members and resource managers are that participatory approaches, underpinned by a commitment to dialogue, are powerful means to promote democratic engagement in environmental management. In both cases, the deliberation and reconciliation of diverse values and perspectives enabled building equitable relationships between project participants. These cases both demonstrate how participation in resource management can be advanced dialogically through community-driven inclusion of values-based, nonexpert, nontechnical, and indigenous voices alongside those of scientists and government agency representatives. This kind of multi-perspective dialogue can clearly assist in the development of governance models that facilitate integrated management and the realization of shared goals. Such governance models become imperative as the drive toward more inclusive participatory processes gains momentum. Nonetheless, as these cases have shown, practitioners must bear in mind that "participation" is never a simple formula with which to address complex, frequently locally particular problems. Nor is a "one size fits all" general theory of participatory practice available. Public participation is never absent of a range of participant positions, of relative power relationships among participants, nor absent a structural, regulatory, or institutional context that organizes participation in particular ways.

Like Flyvbjerg (2001), our ultimate concern is, therefore, power. How power shapes deliberative process and promotes or hinders moves toward collaborative and integrated management is a crucial issue. Certainly stakeholders in both cases are willing to participate and deliberate under conditions in which conventional power relations (which give greater environmental governance authority to government agencies than to citizens) are either minimized or reversed (through indigenous leadership). This setting aside of conventional power disparities proved to be only temporary in the case of Mimiwhangata and potentially only temporary in the case of IKHMG. Nonetheless, critically important for those designing and implementing participatory projects is the development of participatory process ways to illuminate, understand, and manage power disparities among participants. As we have demonstrated, the phronetic interest in practical considerations must focus practitioner attention on local stakeholder dynamics and the institutions that locate specific instances of participatory action. The building of trusting, dialogic relationships, of shared visons and goals; the inclusion of diverse rationalities; and the recognition and negotiation of power relations are all critical steps, guided by phronetic method and inquiry. Inevitably, too, the phronetic method guides practitioners and participants to question and improve the regulatory or institutional regimes that shape environmental decision making and the "participatory infrastructure" available to citizens.

Ultimately, questions of authority, governance control, and regulation cannot be directly addressed meaningfully in the absence of a statutory framework in which to ground integrated or collaborative innovations in environmental management.[16] Indeed, the extent to which innovations in participation, dialogue, and governance models happen should not be left to the goodwill and flexibility of participants. The need for participatory models is particularly the case in relation to ecosystems that span political or regulatory control, or in which moves toward postcolonial structures of governance are implicated. To resolve these questions, current statutory frameworks must include provisions that facilitate participatory decision-making processes. Along with supporting community-led deliberative process, governance attention must also be focused on national/regional policy development and how this can produce innovative resource and environmental management.

Without frameworks that create governance models underpinned by inclusive participatory decision making, collaborative approaches are under-mined and may be viewed as tokenistic. This is especially the case where Maori concepts and values are excluded. Citizens—Maori and non-Maori—may then lose motivation and become unwilling to engage. This risks the loss of community engagement with and support of environmental initiatives, which is now considered an essential ingredient of successful initiatives (Bell, Hampshire, & Tonder, 2008; Campbell & Vainio-Mattila, 2003).

This chapter has illustrated the need to examine current modes of participation and dialogue in environmental management. The potential does exist for participatory initiatives to be translated into equitable forms of governance and management; however, the role played by power rela-tionships in structuring these initiatives requires sustained investigation. We argue that a phronetic approach is valuable, with its focus on the deliberation of values and knowledges, on practice, and on the importance of examining power relationships. The cases discussed here illustrate how communities are able to drive processes through which "practical wisdom" and the "polyphony of voice[s]" (Flyvbjerg, 2001) are collected and combined. At the same time, values and priorities are discussed and negotiated (rather than idealized), thus upholding and expanding participatory norms. This is critical when indigenous voices are embedded in the dialogue. We risk, however, undermining environmental and social gains generated by high-quality participation unless such processes are underpinned by a supportive legislative framework.

Notes

1. Here power is productive and positive as well as dominating. It is viewed as a web of omnipresent relations, which are dynamic within the network of social relationships. The central question is how power is exercised, and careful analysis of the power dynamics of specific practices is a core concern (adapted from Flyvbjerg, 2004, p. 293).

2. Tangata whenua (people of the land) refers to indigenous New Zealanders, the Māori.

3. Flyvbjerg's phronetic social science is embraced by political scientists within the *Perestroika* movement. This movement challenges quantitative approaches within political science focused on developing theory, arguing that political science should focus on practical issues that matter to people rather than on developing complicated theories. See Schram and Caterino (2006).

4. Voices here taken to mean voice, perspective, knowledge, values, and worldview (Senecah, 2004; Flyvbjerg, 2001).

5. Within this strategic framework for water management, the legislative implications are still being considered, and the need for legislative amendment is not yet resolved. The Canterbury Water Management Strategy has been developed to address emerging issues of resource allocation and water quality under conditions of use intensification and declining water quality (CWMS, 2009).

6. The statutory framework in New Zealand is centered on the Resource Management Act 1991 (RMA 1991), which requires the articulation of national policy statements and environmental standards at a national level, with land use planning and regulation devolved to local and city councils.

7. RMA 1991 Sec. 36B provides for "joint management agreements" between local governments, statutory authorities, the Crown, and iwi.

8. The Waitangi Tribunal is a commission of inquiry charged with investigating both historic and contemporary breeches of the Treaty of Waitangi (1840) and issues nonbinding reports on claims brought to it by Māori claimants.

9. These priority areas are biodiversity, sustainable fisheries, restoring and protecting the mauri (lifeforce), addressing climate change, socioeconomic issues, and integrated management.

10. See, for instance, Te Uri o Hau Settlement Act of 2002.

11. Regulatory authority for the harbor is spread across two district councils, one regional council, one unitary authority, and several central government agencies.

12. For an account of this event, see IKHMG (2015).

13. Other stakeholders were involved in this process also, but were peripheral to the core partnership. These include Ngatiwai Trust Board, local government authorities, other government departments (for example, the NZ Ministry of Fisheries), and other community groups.

14. Limited customary management of fisheries of this nature is possible through the *Fisheries Act 1996*.

15. Especially because of the historic context in which Maori perspectives have often been marginalized.

16. See CWMS (2009).

References

Adger, W. N., Brown, K., & Tompkins, E. L. (2005). The political economy of cross-scale networks in resource co-management. *Ecology and Society, 10*(2), 9.

Anderies, J. M., Walker, B. H., & Kinzig, A. P. (2006). Fifteen weddings and a funeral: case studies and resilience-based management. *Ecology and Society, 11*(1), 21.

Armitage, D. (2005). Adaptive capacity and community-based natural resource management. *Environmental Management. 35*(6), 703–715.

Ballantine, W. J., Grace, R. V., & Doak, W. (1973). *Mimiwhangata 1973 report*. Unpublished report to Turbott and Halstead, Auckland. Held at University of Auckland Leigh Marine Laboratory.

Bell, S., Hampshire, K., & Tonder, M. (2008). Person, place, and knowledge in the conservation of the Saimaa ringed seal. *Society & Natural Resources, 21*(4), 277–293.

Berkes, F. (2008). The evolution of co-management, *Journal of Environmental Management, 90*, 1692–1702.

Berkes, F., Colding, J., & Folke, C. (Eds.). (2003). *Navigating social-ecological systems: Building resilience for complexity and change*. Cambridge, UK: Cambridge University Press.

Brulle, R. J. (2010). From environmental campaigns to advancing public dialogue: Environmental communication for civic engagement. *Environmental Communication, 4*(1), 82–98.

198   Dodson, Palliser

Campbell, L. M., & Vainio-Mattila, A. (2003). Participatory development and community-based conservation: Opportunities missed for lessons learned? *Human Ecology, 31*(3), 417–437.

Canterbury Water Management Strategy (CWMS). (2009). Retrieved from http://ecan.govt.nz/publications/Plans/cw-canterbury-water-wanagement-strategy-05-11-09.pdf

Cocklin, C., Craw, M., & Mcauley, I. (1998). Marine reserves in New Zealand: use rights, public attitudes, and social impacts. *Coastal Management, 26*(3), 213–231.

Conley, A., & Moote, M. (2003). Evaluating collaborative natural resource management. *Society and Natural Resources, 16,* 371–386.

Coombes, B., & Hill, S. (2005). "Na whenua, na Tuhoe. Ko D.o.C te partner"— Prospects for co-management of Te Urewera National Park. *Society and Natural Resources, 18*: 135–152.

Cox, R. (2007). Nature's "crisis disciplines": Does environmental communication have an ethical duty? *Environmental Communication, 1*(1), 5–20.

Cundill, G., Cumming, G. S., Briggs, D., & Fabricius, C. (2012). Soft systems thinking and

social learning for adaptive management. *Conservation Biology, 26*(1), 13–20.

Department of Conservation (2004). *Marine reserve proposal: Mimiwhangata Community discussion document.* Whangārei: Department of Conservation.

Department of Conservation (DOC). (2005). *MPA policy and implementation plan.* Wellington: Department of Conservation.

Dodson, G. (2014a). Co-governance and local empowerment? Conservation partnership frameworks and marine protection at Mimiwhangata, New Zealand. *Society and Natural Resources, 27*(5), 521–539.

Dodson, G. (2014a). Moving forward keeping the past in front of us: Treaty Settlements, conservation, co-governance and communication, in G. Dodson & E. Papoutsaki (Eds.), *Communication issues in Aotearoa New Zealand* (pp. 62–73). Auckland: Unitec ePress.

Durie, M. (1998). *Te Mana, Te Kawanatanga; The politics of Maori self-determination.* Auckland: Oxford University Press.

Dutta, M. (2011). *Communicating social change: Structure, culture, agency.* New York: Routledge.

Flyvbjerg, B. (1998). *Rationality and power: Democracy and practice.* Chicago: University of Chicago Press.

Flyvbjerg, B. (2001). *Making social science matter: Why social inquiry fails and how it can succeed again.* Cambridge: Cambridge University Press.

Flyvbjerg, B. (2004). Phronetic planning research: Theoretical and methodological reflections. *Planning Theory & Practice, 5*(3), 283–306.

Flyvbjerg, B. (2005). Social science that matters. *Foresight Europe, 2,* 38–42.

Flyvbjerg, B. (n.d.). *What is phronetic planning? What is phronetic social science?* Retrieved from http://flyvbjerg.plan.aau.dk/

Forster, M. (2010). Recovering our ancestral landscapes: A wetland's story. In R. Selby, P. Moore, & M. Mulholland (Eds.), *Māori and the Environment: Kaitiaki* (pp. 199–218). Wellington: Huia Publishers.

Funtowicz, S., & Ravetz, J. (2003). Post-normal science. *International Society for Ecological Economics (ed.), Online Encyclopedia of Ecological Economics.* Retrieved from http://www. ecoeco. org/publica/encyc.htm

Funtowicz, S., & Ravetz, J. R. (2003). Post-normal science. International Society for Ecological Economics: *Internet encyclopedia of ecological economics.* Retrieved from http://isecoeco.org/pdf/pstnormsc.pdf

Grace, R. V. (1981). *Paparahi marine survey.* Archival document, Nga Maunga ki te Moana Trust: New Zealand.

Hagendijk, R., & Irwin, M. (2006). Public deliberation and governance: Engaging with science and technology in contemporary Europe. *Minerva, 44,* 167–184.

Hargreaves, T. (2012). Questioning the virtues of pro-environmental behaviour research: towards a phronetic approach. *Geoforum, 43*: 315–324.

Harmsworth, G., & Awatere, S. (2013). Indigenous Māori knowledge and perspectives on ecosystems. In Dymond, J. (Ed.), *Ecosystem services in New Zealand: Conditions and trends* (pp. 274–286). Lincoln, NZ: Manaaki Whenua Press.

Healy, S. (2009). Toward an epistemology of public participation. *Journal of Environmental management, 90*(4), 1644–1654.

Horsbol, A., & Larssen, I. (2012). Public engagement as a field of tension between bottom-up and top-down strategies: Critical discourse moments in an "Energy Town." in L. J. Phillips, A. Carvalho & J. Doyle (Eds.), *Citizen voices: performing public participation in science and environment communication* (pp. 163–187). Bristol, UK: Intellect.

Integrated Kaipara Harbour Management Group (IKHMG). (2011). *Integrated Strategic plan of action.* Retrieved from http://www.kaiparaharbour.net.nz/ Publications

Integrated Kaipara Harbour Management Group (IKHMG). (2015). *IKHMG Symposium, 2014.* Retrieved from https://www.youtube.com/watch?v=iBy8P1eKvcs

Kawharu, M. (2000). Kaitiakitanga. *Journal of the Polynesian Society, 109,* 349–370.

Land and Water Forum (2010). *Report of the Land and Water Forum: A Fresh Start for Fresh Water.* Retrieved from http://www.landandwater.org.nz/Site/Resources.aspx

Land and Water Forum (2012a). *Second report of the Land and Water Forum: Setting limits for water quality and quantity, and fresh water policy and plan making through collaboration.* Retrieved from http://www.landandwater.org.nz/Site/Resources.aspx

Land and Water Forum (2012b). *Third report of the Land and Water Forum: Managing water quality and allocating water.* Retrieved from http://www.landandwater. org.nz/Site/Resources.aspx

Marsden, M. (1992). *Kaitiakitanga: A definitive introduction to the holistic worldview of the Māori.* Retrieved from http://www.marinenz.org.nz/documents/ Marsden_1992_Kaitiakitanga.pdf

McCallum, W., Hughey, K. F., & Rixecker, S.S. (2007). Community environmental management in New Zealand: Exploring the realities in the metaphor. *Society and Natural Resources. 20*(4): 323–336.

Ministry for Primary Industries (2015). *West Coast South Island: Proposed marine protected areas (using fisheries regulations) decision document.* MPI Discussion Paper No. 2015/28. Prepared for the Minister for Primary Industries by the Ministry for Primary Industries. ISBN No: 978-0-908334-85-8 (online) ISSN No: 2253–3907 (online).

Moller, H., Lyver, P., Bragg, C., Newman, J., Clucas, R., Fletcher, D., Kitson, J., McKechnie, S., Scott, D., & Rakiura Titi Islands Administering Body (2009). Guidelines for participatory action research partnerships. *New Zealand Journal of Zoology, 36,* 211–241.

Mutu, M. (2010). Ngāti Kahu kaitiakitanga. In R. Selby, P. Moore, & M. Mulholland (Eds.), *Māori and the environment: Kaitiaki* (pp. 13–36). Wellington: Huia Publishers.

Ostrom, E., Janssen, M. A., & Anderies, J. M. (2007). Going beyond panaceas. *PNAS, 104*(390): 15176–15178.

Phillips, L. (2011). *The promise of dialogue.* Amsterdam: John Benjamins BV.

Phillips, L. J., Carvalho, A., & Doyle, J. (2012). *Citizen voices: performing public participation in science and environment communication.* Bristol, UK: Intellect.

Plummer, R., & Armitage, D. (2010). Integrating perspectives on adaptive capacity and environmental governance. In D. Armitage & R. Plummer (Eds.), *Adaptive capacity and environmental governance* (pp. 1–19). Heidelberg, Dordrecht, London, New York: Springer.

Popa, F., Guillermin, M., & Dedeurwaerdere, T. (2015). A pragmatist approach to transdisciplinarity in sustainability research: from complex systems theory to reflexive science. *Futures. 65,* 45–56.

Reed, M. S. (2008). Stakeholder participation for environmental management: a literature review. *Biological Conservation, 141,* 2417–2431.

Rotarangi, S., & Russell, D. (2009). Socio-ecological resilience thinking: Can indigenous culture guide environmental management? *JRSNZ, 39,* 4, 209–213.

Russ, G. R., Alcala, A. C., Maypa, A. P., Calumpong, H. P., & White, A. T. (2003). Marine reserve benefits local fisheries. *Ecological Applications, 14*(2), 597–606.

Schram, S. F., & Caterino, B. (2006). *Making political science matter: Debating knowledge, research, and method.* New York: New York University Press.

Senecah, S. (2004). The trinity of voice: The role of practical theory in planning and evaluating the effectiveness of environmental participatory process. In S. W. Depoe, J. W. Delicath, & M. F. Aepli Eisenbeer (Eds.), *Communication and public participation in environmental decision-making* (pp. 13–33). Albany: State University of New York Press.

Servaes, J. (2008). *Communication for development and social change.* Thousand Oaks, CA: Sage.

Singhal, A. (2001). *Facilitating community participation through communication.* New York: UNICEF.

Tipa, G. (2010). Cultural opportunity assessments: Introducing a framework for assessing the suitability of stream flow from a cultural perspective. In R. Selby, P. Moore, & M. Mulholland (Eds.), *Māori and the Environment: Kaitiaki* (pp. 155–173). Wellington: Huia Publishers.

Tipa, G., & Welch, R. (2006). Co-management of natural resources: issues of definition from an indigenous community perspective. *The Journal of Applied Behavioural Science, 42*(3), 373–391.

Tūhoe Deed of Settlement (2012). Retrieved from http://nz01.terabyte.co.nz/ots/ fb.asp?url=livearticle.asp?ArtID=363804612

Ulrich, W., & Reynolds, M. (2010). Critical systems heuristics. In M. Reynolds & S. Holwell (Eds.), *Systems approaches to managing change: A practical guide* (pp. 243–292). London, UK: Springer.

Uunila, L. (2003). Community involvement in New Zealand marine reserve management: Examining practice. In *Science and stewardship to protect and sustain wilderness values,* Seventh World Wilderness Congress Symposium, 2001 November 2–8, Port Elizabeth, South Africa (pp. 142–147). Co-sponsored by U.S. Department of Agriculture, Forest Service, Rocky Mountain Research Station.

Waikato River Authority (2013). *Annual report 2013,* Hamilton: Waikato River Authority.

Tribunal, W. (2011). Ko Aotearoa tēnei: A report into claims concerning New Zealand law and policy affecting Māori culture and identity. *Waitangi Tribunal, Wellington.*

Walker, G. (2007). Public participation as participatory communication in environmental policy decision-making: From concepts to structured conversations. *Environmental Communication, 1*(1), 99–110.

Walker, G. B., Senecah, S. L., & Daniels, S. E. (2006). From the forest to the river: Citizens' views of stakeholder engagement. *Human Ecology, 13*(2), 194–202.

Warner, J. (2007). *Multi-stakeholder platforms for integrated water management.* Aldershot: Ashgate Publishing.

Chapter 8

Toward Communicative Space

A Maritime Agora of Backrooms and Thoroughfares

CHUI-LING TAM

Introduction

Participation is contested, viewed variously as tyrannical (Cooke & Kothari, 2001) and transformative (Hickey & Mohan, 2004). The United Kingdom's respected Institute of Development Studies (2016) calls participation "a political endeavour that challenges oppression and discrimination, in particular of the poorest and most marginalised people." Albeit flawed, participation is seen as critical to enacting democracy and solving environmental challenges through active, engaged citizens affected by or otherwise interested in the outcomes of decision-making processes (Hurlbert & Gupta, 2015; Turnhout, Van Bommel, & Aarts, 2010). Participation commands a varied basket of grassroots strategies, such as public protest, whereas public participation tends to be a deliberative exercise designed by authorities wherein citizens are invited to take part (Cornwall & Coelho, 2007). This is an unresolved tension in participation with sociospatial implications for participation design, practice, and assessment that can inform and move the field closer to effective community engagement in contextually diverse locales.

Views differ on the objectives and design of public participation, but generally it can be defined as normative as it aims to influence decisions, enhance democratic capacity, advance social learning, and empower marginalized individuals and groups, coupled with a substantive task to mobilize diverse information and knowledge, with instrumental goals of generating legitimacy and resolving conflict (Glucker, Driessen, Kolhoff, & Runhaar, 2013). Good public participation promotes citizenship, knowledge sharing, and conflict resolution; these combined aims ideally increase the appropriateness of policy decisions, the probability of community support and compliance, and therefore the likelihood of successful implementation.

A chief aim of public participation is the creation—often implicit—of spaces for inclusive communication. Such "communicative spaces" include both the physical setting in which participants meet and the social composition of such gatherings. Public participation is central to enacting citizenship and the communicative dimensions of social, cultural, and political rights within the public sphere (Gomez, 2011) as a communicative space. Whether in environmental management or other undertakings such as disease prevention, sound community organization and participation depend on the creation of communicative spaces wherein participants can identify problems and solutions to a given problem; these are both material and discursive spaces where people can have a voice (R. de Souza, 2009).

Participation is an appealing trope to empower the oppressed, and therein lies the tension in this chapter. The taken-for-granted nature of participation must be challenged by scrutinizing the specific conditions and presuppositions—not always benign—that determine the shape and framing of participation before it reaches the public sphere (Huxley, 2013). Participation is often designed as publicly accessible events scheduled to take place at designated sites and times. These include task forces, royal commissions, workshops, public consultations, town halls, and focus groups through which citizens can ostensibly share their thoughts, experiences, and grievances. Such deliberative arenas serve invited participation, but participation also occurs in informal arenas where social movements grow out of citizens' grassroots interaction (Cornwall & Coelho, 2007). Participation depends, in part, on a democratic process in which the force of the better argument prevails and the exercise of individual agency realizes communicative ideals of open and truthful engagement in debate (Habermas, 1984, 1987). Participation is also enacted by actors sharing space physically and virtually as advanced communication technologies such as the Internet "conquer" the distancing effects of time-space (Janelle, 1968; Torrens, 2008).

Despite decades of efforts to engage communities in resource management, managers, policy makers, researchers and nongovernmental organizations (NGOs) repeatedly fail to involve stakeholders in trustful and effective communication (see Bennett & Dearden, 2014; Chuenpagdee et al., 2013; Elliot, Mitchell, Wiltshire, Manan, & Wismer, 2001; Marques, Ramos, Caeiro,

& Costa, 2013; Oyanedel, Marín, Castilla, & Gelcich, 2016; Pullin et al., 2013). Communication failure occurs in a participation context not unlike an agora, which serves as a social space of shared ideas and as a physical space where communication occurs (Bryant, 2006). When public participation fails to engage communities, participatory processes and communication strategies warrant investigation. Chuenpagdee et al. (2013, p. 234) suggest that failure occurs when marine protected areas (MPAs)—and I would add resource management regimes generally—"do not deliver what they intend to do," prompting stakeholders to lose faith in the regime. Management failure occurs for various reasons, such as the poor quality of project design, unqualified actors entrusted with implementation, and low transfer of power and authority to marginalized actors (Dyer et al., 2014). However, the point here is this: we need to scrutinize the communication environment as a specific condition of participation. Attention to the social, spatial, and temporal dimensions of human communication at sites of resource conflict would arguably enhance the design of participation to ensure marginalized voices are heard and heeded, thereby increasing the likelihood of locally appropriate strategies that produce substantive environmental outcomes.

These three dimensions intersect uneasily with resource management interventions in the Tiworo Straits and Wakatobi National Park (WNP) in Indonesia. In these adjacent fisheries in Southeast Sulawesi province, public participation is at odds with the communication cultures of local maritime environments, thereby marginalizing maritime space and peoples. Here, marine dwellers engage in diverse conversations in public walkways and private abodes left out of deliberative, invited participation. Such communicative spaces form a maritime agora of backrooms and thoroughfares where information, aspirations, challenges, and relationships are discussed, demonstrated, and performed. The agora is, to adapt Cons (2015), a zone of engagement, an interstitial or *between* space occupied by citizens who may be bypassed in public participation. Recognizing the existence of such spaces, and accounting for how, what, and which publics communicate in such spaces, is arguably critical to good participation.

This chapter contributes to research in communication geography, which is experiencing a contemporaneous "spatial turn" in communication and "communicative turn" in geography (Adams & Jansson, 2012). It interrogates the spatial particularity of communication practice, environments, and voice and its implications for participation infrastructures. The chapter begins with a review of public space and participation, followed by two studies of marine intervention, participation, and communication space: a stream fish pond project in Tiworo and public consultations during the rezoning of WNP. In both initiatives, the government bodies and NGOs involved had, by their own reckoning, engaged in participatory, community-based management. By contrast, many villagers told the author a different tale of miscommunication and exclusion. These are tales of resource management and communication failure, but also of maritime agoras as communicative, overlooked, *between* spaces that present opportunities for participation. These studies reveal three insights. First, communication chaos and disorder are intrinsic to informal participation. Second, thoroughfares should be viewed not as temporary passing spaces but as habitual time-spaces of engagement that lie between and beyond the sites of deliberative public participation. And third, the boundaries between public and private communicative space are elastic. These insights can enhance environmental communication to promote participation attuned to the diversity of citizens.

Public Space and Participating Publics

Humans "use" environmental goods and services according to their various values of nature (Diehm, 2012; Levy, 2003). Research and practice in public participation, community engagement, and traditional/indigenous/ local ecological knowledge affirm nature as a resource that humans value, extract, protect, share, worship, and experience in complex and conflictual ways (Kellert, 1996; Ku & Zaroff, 2014; van Zyl, 2009). Resources are part of socioecological systems where constant change—in the nature and quality of the resource, in extractive practices, and in the politico-economic context, among others—inspires conflict and uncertainty (Mitchell, 2015). For development and environmental decision makers, participation depends on and creates a communicative space in which diverse actors share ideas, address complexity, and reduce conflict to reach a mutually beneficial outcome (Chambers, 2008; Fortmann, 2008; Reed, 2008). However, participation designed from above by authorities laboring under non-local presuppositions, as is quite common in Indonesia, risks generating low levels of influence on decision making, low resource access, and ultimately fewer benefits for the marginalized (Glaser, Baitoningsih, Ferse, Neil, & Deswandi, 2010).

AGORAS

Notions of communicative space are deeply embedded in the Habermasian ideal of a liberal democratic public sphere in which actors come together with a commitment to communicate truthfully and to consent to be governed by that democratic process (Plot, 2009), presumably in a space of social and physical security. Habermasian public space embodies Western norms—inherited from the European Enlightenment—as the intermediary sphere between civil society and the state, a physical space accessible to all citizens, and an abstract notion of a politically informed and engaged public (Touaf & Boutkhil, 2008). Habermasian idealists view citizens as emboldened and informed actors who exercise the communicative right, the communicative will, and the communicative responsibility to engage in debate on matters of public importance such as resource management. However, this ideal is contested. In the Western urban planning contexts where Habermas's ideal has held most sway in the past two decades, skeptics argue it fails to account for power that interferes with open, citizen-empowering public participation (Fischler, 2000; Flyvbjerg, 1998; Flyvbjerg & Richardson, 2002; Sager, 2013). The Habermasian citizen is not culturally universal. Communication can be fraught with cultural, spatial, and political particularities that compromise good participation in environmental decision making, as will be seen in the Indonesian spaces presented here.

The idea of public space can be traced further back to the agoras of Ancient Greece, where agoras were central places in city-states that reflected the democratic ideals of an orderly, engaged citizenry (Geoghegan & Powell, 2009). The agora was a public space wherein citizens could participate in matters of collective and individual concern. It was a specific physical space of informal public assembly, serving as an open-air antechamber from which citizens then walked to the city-state's formal religious center or assembly place where decisions were legislated. The agora was also a conceptual and hybrid space, both private and public (Powell, 2012). In the modern world, the agora can be a physical space such as a coffee shop, a Speaker's Corner, a community center temporarily used for planning consultations, a public hearing on an oil pipeline or fishery quotas, or any setting or "social action context" (Jenlink & Banathy, 2002, p. 469) in which civil society can make collective decisions. In such settings, public engagement is an ever-present possibility untethered to any specific place or time.

Agoras are discursive spaces. For instance, media agoras encompass both the mainstream media of public, commercial media organizations and professionals, as well as the many communication tools that enable participation and public representation (Garcia-Blanco, Van Bauwel, & Cammaerts, 2009). Participation is enabled by "new agoras" such as digital space (Bryant, 2006; Iyall Smith, 2007; Jenlink & Banathy, 2002) where social media, virtual contact, and electronic surveys enable citizens to exercise voice covertly. Participation in environmental decision making can be viewed as "science-society agoras" with a three-part framework: first, it is a forum for debate open to all citizens; second, all citizens must have the opportunity, even if not taken up, to actively participate in debate; and third, interactivity is possible whereby citizens can exchange views and challenge each other on issues raised (Davenport & Leitch, 2005).

Agoras are dynamic spaces that evolve beyond their original plan as people perform their individual and group identities (Touaf & Boutkhil, 2008). Across history, agoras were used informally as spaces of commerce and trade, a marketplace of material and social exchange of goods, ideas, and gossip; and zones of engagement among citizens producing and contesting representations of their identities, good governance, and societal values. The agora serves as a social and communal public-private sphere with physical, abstract, and digital dimensions. It is both an informal and a deliberative space in which diverse activities and interactions may occur simultaneously, mutating with the nature of activities and interactions that occur at different times. The agora exemplifies communicative space where individuals exercise collective power by gathering openly to converse, to engage in social life, and to share stories.

However, the agora is also a constrained space. As a political resource, it is neither accessed by nor accessible to all. While Habermasian public space is grounded in a specific European historical time and space of engagement, the average citizen in a developing country today does not identify public space as essential to good governance (Touaf & Boutkhil, 2008). Nor was the Ancient Greek agora a space of free assembly for all persons, for only Greek males had the legal status of citizens; slaves, women, and other lesser persons had no voice in the agora (Davenport & Leitch, 2005). The agora, then, has a contradictory history as a space of inclusion and exclusion. This is reproduced in modern public space, where social practice and expectations, discourses, or zoning regulations have the effect of excluding undesirable peoples or behaviors. For instance, in Morocco, the emergence of women contraband traders has altered the cityscape and increased participation in illicit activity, but also disrupted the traditionally male public sphere of economic production (Touaf & Boutkhil, 2008). In forest communities around the world, women's claims to resource use rights have not only physically placed them in resource spaces controlled by men, but also increased their confidence, assertiveness, and political and economic participation in traditionally male-dominated public spheres (Coleman &

Mwangi, 2013); their entry into the public sphere of forest associations tended to reduce the likelihood of disruptive conflict because they were more able to work collaboratively in keeping with the normative, substantive, and instrumental goals of participation. Lastly, Habermasian communicators are presumed to have the self-awareness to engage truthfully in debate, but many people fail to recognize their own privilege or that communication is a non-neutral phenomenon (Rienstra & Hook, 2006). Communication can serve to democratize public engagement, but it also shapes politicoeconomic relations through the spread of ideas, consumer tastes, and market demand that foster unequal economic and discursive power (Innis, 1986; Tremblay, 2012). This becomes significant in the two case study sites because they are situated within global chains of fishery commodity production and global discourses of conservation and development.

Participatory Methodology

Participation captures a broad range of methods, settings, and problems; it has been particularly popular in the fields of development, environmental and urban studies, youth, and community health. While concepts of participation have been scaled up to capture diverse populations responding to national and global concerns such as climate change, conservation, and sustainable development (Castells, 2008; Cox, 2013; Powell, 2012), participation remains attached to its local, rural, community roots as a paradigm that lifts up and includes disadvantaged people who live immediately with the effects of policy decisions (Chambers, 2008; Freire, 1970). The breadth of participatory methodology is largely derived from participatory rural appraisal (PRA) and associated approaches such as participatory action research (PAR) and agroecosystem analysis (Cornwall & Pratt, 2011; H. N. de Souza et al., 2012).

Four key features of participatory methodology are germane to agoras and the publics who inhabit them. The first is the basket of methods that can be deployed to promote the participation of diverse publics—especially the poor and otherwise marginalized. The second feature is the attitude and behavior of outside specialists and powerful others who enter the local lived spaces of identified participants. The third feature is the explicit acknowledgment that marginal publics have worthwhile knowledge of their own circumstances and can be the authors of their own emancipation. And the fourth feature is an emphasis on informal communication in nonformal settings allowing poor people's knowledge to flow upward to those in power and among resource users with potentially conflictual priorities. These four features comingle in a spatial politics—rarely explicitly acknowledged—that is central to the role reversals required of participation. Indeed, Chambers (1997, p. 1750) invokes space when he sharply advocates for a "pedagogy for the non-oppressed" whereby privileged, educated elites are required to complement their institutional training with a week of immersion in a village or a slum. The message here is that public participation needs to expand beyond the orderly Habermasian ideal to more actively include the diversity of available participation methodologies and to further identify informal communication space to improve the design of participation exercises.

It is clear that participation requires an engagement with place-bound publics, regardless of the scale of the place or the public. While modern conceptions of the agora have expanded into the virtual realm and beyond the physical constraints of time-space, space remains a tenacious consideration within participation in three key ways. First, participation can be designed such that it invites the public into a designated space, usually as part of a group exercise to facilitate overt information sharing; focus groups and public workshops are typical of this model. Second, participation can be conducted remotely across time-space through use of media technology such as mailed questionnaires, telephone interviews, the Internet, or other cyber forums. Third, participation can be performed as a direct one-on-one inter-action between parties in a shared space with invested time; this is useful in cross-cultural contexts or among strangers where corporeal co-presence helps reduce social distance (Bowlby, 2012; Tam, 2015). Whatever the format of participation, spatial choices affect its design.

The spatiality of participation becomes particularly problematic in resource management areas with disperse populations; this is often true of marine environments, which face multiple pressures related to development, population growth, and environmental change (Chuenpagdee et al., 2013; Neumann, Vafeidis, Zimmermann, & Nicholls, 2015; Tam, 2015). To man-age the time and cost constraints of traversing space to achieve focused and effective communication, public participation is conducted as a deliberative process with orderly, democratic rules of engagement such as an agenda, turn taking, and time allocation. Deliberative participation often involves a managed schedule of spaces to facilitate a communicative activity such as focus groups, surveys, and town hall meetings; it is purpose driven, and whose purpose is driving the activity matters. Such participatory exercises were commonly used in the two Indonesian sites to be discussed. Deeper and more creative local engagement might involve community mapping, participatory modeling, and theater; these group activities may be conducted in locations such as in a forest clearing, a neighborhood square, or a villager's house (Abah, 2007; Yates, 2014).

COMMUNICATION CHAOS AND COMPLEXITY

Communication is complex and may not fit deliberative norms. Agoras accommodate deliberative and informal participation, imbued with the characteristics and challenges of their particular social, spatial, and temporal context. Agoras raise questions about the communication geography of resource management processes such as intervention, informing, knowledge sharing, decision making, implementation, and participation. While the Habermasian agora is often a deliberative space of scheduled meetings, published agendas, turn taking, and orderly seating, there are other agoras where the messiness of communication can yield fruitful dialogue.

In these other agoras, multiple disorderly interactions can happen simultaneously. Such agoras are chaotic. This has resonance with informal communication in participatory methodology, which relies on a basket of methods to promote the participation of diverse and differently marginalized publics (Chambers, 2008). Outside specialists and powerful others who enter the communities facing regime change need to demonstrate a communicative attitude by taking part in these chaotic arenas. Such behavior can reinforce acknowledgment that marginal publics have worthwhile knowledge to communicate in their own way and their own agoras.

Communication travels between senders and receivers in multidirectional networks of dissemination (Rogers & Kincaid, 1981). Given that public participation tends to designate points of communication, it risks overlooking the space *between* such points. Recognizing *between* spaces, and accounting for how, what, and which publics communicate in such spaces, is critical for good participation, especially in contexts where local engagement, social hierarchies, research methodologies, and inadequate communication and transportation technologies may present barriers to people accessing participatory exercises (Tam, 2006, 2015).

Informal communication can accommodate the accidental nondeliberative stories as people go about their day. The content of communication among a given public—children, for instance—might vary depending on the particular social, spatial, and temporal context; they might talk about different things, at different times, in different places, with different sets of people (Kajubi, Bagger, Katahoire, Kyaddondo, & Whyte, 2014). Communication happens on the street, a dynamic space that assumes diverse functions and identities as people use the street for protest and other political struggles (Kallianos, 2013); the street serves as a mobile space of communicative action. Informal communication is deployed by elites and non-elites, deriving power from local knowledge and place-based networking and thereby influencing, for instance, participatory planning processes (Lee, 2007). Communication and selection are intrinsic to networks, producing effects that may determine stakeholders' readiness to communicate with participants with dissimilar knowledge and values; different phases—that is, *time*—in the management process may also affect the effectiveness of communication (de Nooy & Nooy, 2013).

Formal processes for engaging indigenous peoples are hampered by cross-cultural miscommunication because managers have a poor understanding of local indigenous institutional and social processes, and the effects of community influence and power (Lloyd, Van Nimwegen, & Boyd, 2005). The complexity and challenges of communication are significant, often inspiring wealthy jurisdictions (such as rich countries) to invest institutional, monetary, and personnel resources to improve community engagement. But what might be the communication-related outcomes of participation in poor countries with limited resources to invest? This is worth asking in the context of the underdeveloped communities of the Indonesian littoral.

The extant literature suggests there is room for deeper engagement with communication space. The task of this chapter is to conceptualize the arena of resource management interaction as a maritime agora composed of both public and private space, and contribute to the spatial-communicative turn in environmental communication, resource management, and marine policy.

A Tale of Two Fisheries

Indonesia is the world's largest archipelago with the fourth-largest human population dispersed across its 17,500 islands. The maritime environment is critical to human settlement in Indonesia and indeed throughout archipelagic Southeast Asia. Ongoing degradation of marine species health, habitats, and biodiversity pose a chronic challenge. In Southeast Sulawesi province, MPAs, aquaculture, tourism, new value-added marine products, economic resource diversification, and partnerships with outside interests such as environmental NGOs (ENGOs) are among innovations adopted to reduce marine degradation. Similar interventions have occurred in the Tiworo Straits and in Wakatobi, two underdeveloped areas that are significant contributors to provincial wealth, economic livelihoods, and personal well-being. Both fisheries are under threat, and both are responding to environmental change in ways that raise questions about the intersection of public space, communication, and changing resource management regimes. In Tiworo, one response was a stream fish pond project in Lasama village. In Wakatobi, the response was an MPA, which has deeply affected Sama Bahari village. For local inhabitants, the success of these initiatives is debatable.

These are populated but remote, infrastructure-challenged spaces. From the provincial capital of Kendari on the Southeast Sulawesi mainland, the cheapest public marine transportation is a morning ferry heading southeast for three hours to the large island of Muna, then continuing for another two hours to Buton island. From here, travelers can while away the daytime hours before taking a 12-hour night ferry to Wakatobi. Next, travelers traverse a series of paved and unpaved roads and water bodies to reach their final destination. Electrical power in villages is only available for a few hours at night. In 2002, there were no telephones in villages. Today, cellphones are common, and texting is ubiquitous; however, cellular networks are unreliable, Internet access is very limited, and home computers are beyond the means of most villagers.

The Tiworo Straits lie between the Southeast Sulawesi mainland and Muna. Lasama is a coastal village on the south shore of the Tiworo Straits, supporting about 700 residents at the time of study in 2002. It has a multi-ethnic population closely integrated through family and commercial networks and living in close proximity near the main road leading south to larger settlements. Villagers identify primarily as fisherfolk, with some pursuing limited agrarian livelihoods. The current analysis focuses on Lasama and the "community-based" stream fish pond project, called the *Empang Parit*, that was partially constructed there but never completed. Many of Lasama's fishing residents had lived on small islands in the Tiworo Straits before a government-ordered resettlement to the coast to ease population pressure on the islands; they adapted their livelihoods to the increased distance to traditional fishing grounds, a declining fishery resource, and new economic opportunities. The author lived with various families in the village for three months, achieving Indonesian language proficiency and a familiarity with local customs and relationships that helped establish her later credibility and guided her interactions and observations within marine and coastal communities around Southeast Sulawesi.

Wakatobi National Park (WNP) was established in 1996 and is composed of four large main islands supporting 100,000 inhabitants, most of whom engage in fishing livelihoods; some work as farmers, tourism operators, civil servants, and shopkeepers, among others. The author conducted interviews and observation across 15 WNP villages during 2011 and 2013, but the current analysis focuses on Sama Bahari, which is almost exclusively populated by some 2,000 ethnic Bajau. The village is built atop a coral atoll. As a fairly homogenous village, the people favor speech in ethnic Bajau, although most are fluent in the official Indonesian.

At both sites, livelihoods depend on harvesting reef fish including the now protected and lucrative Napoleon wrasse, pelagic species such as tuna and shark, sea cucumber, and shellfish such as mollusks, abalone, and crab. Tiworo and WNP have experienced harmful extractive practices such as cyanide and blast fishing (which has reputedly declined) and an expanding use of fish fences to compensate for the declining size of catch and individual fish caught. While some fishermen may earn IDR30,000 (equivalent to about two U.S. dollars) per day, the more productive fishermen can more than three times that figure depending on equipment used and time invested.

Marine dwellers for the most part support the principles of sustainable development and agree with the need for innovative strategies and coordination to conserve their deteriorating marine resource. Where they differ is on strategies to achieve it, as well as the distribution of harm and benefit. The resource management strategies—the stream fish pond in Tiworo and the internationally acclaimed WNP—were presented to the respective communities as sustainable development initiatives that would alleviate poverty and reduce resource degradation. Both sites exemplify externally driven development and environmental communication that failed to accommodate the norms in local maritime agoras. Rather, participation and community-based management were key tropes invoked by policy makers and NGOs to cultivate villagers' compliance and to publicize development efforts ostensibly in keeping with the participation paradigm. Poor communication was entwined with flawed management. In the end, the fish pond was never completed and came to be frequented by illegal mangrove loggers. Meanwhile, WNP underwent two decision-making exercises to establish a multiple-use zoning system without significantly easing local community grievances or confusion.

Discussion

Tiworo and WNP are stories of environmental change and adaptation brought on by outside intervention. They are also stories of participation and implementation failure that did not inspire consistent satisfaction among affected communities. These shortcomings are discussed and analyzed within three themes: communication behavior and content, communication time-space, and communication and management failure.

COMMUNICATION BEHAVIOR AND CONTENT

It is important to understand communication behavior and the context in which it occurs. In deliberative arenas such as public meetings and work-shops, Indonesians emphasize formality, respect for authority, and titles and professional designations that convey expertise. In Tiworo and WNP, these arenas—located at the village hall or more commonly the village chief's house—are in effect communication spaces sanctioned at discreet times by outside authorities intervening in resource management regimes. They are scheduled at fixed times determined by a select few and are often attended by a representative select few citizens. Women are underrepresented or absent and are to be found gossiping in the backroom while preparing refreshments to serve to participants in the front room. In these maritime environments, only select people are communicating in the deliberative space of participation.

There are other communication norms and other agoras. In Tiworo, the locals have a saying: *mari cerita. Cerita* means story, account, narrative, or tale. *Mari* means come hither, or let us (do something). The term is an invitation to gather and tell stories in the immediacy of a shared space and activity. This communication ethos flows through Tiworo and WNP. This is revealed by the research data gathered, but also the act of data gathering as a communication experience. Methods such as random and purposive sampling, transect walks, participation, observation, semistructured inter-views, and casual conversations—and the manner and spaces in which this research communication occurred—reveal insights about public space and their efficacy in fostering communication. The author spoke directly with respondents in Indonesian in the hope of reducing the potential distancing effects of mediated communication through translators. Her research was informed by a communicative methodology.

Trust had to be established with respondents. In Lasama, the author earned trust through field immersion, a demonstrated interest in and empathy with village lives, and patience with women's banter about beauty rituals (which segued into backroom conversations about gender norms that excluded women from the pond project). There were many conversations over laundry by the water wells along the main road, a hybrid space of private activity conducted in public view. In WNP, staff at a respected community NGO helped the author to fast-track trust building by accompanying her to interviews, ready of their own volition to frame her questions into the fishers' ethnic tongue. In one instance, we squatted in the dirt behind a fisherman's house as his wife gutted fish, and a dozen neighbors turned the interview into a heated discussion about government inaction. Research communication was a combination of formal interviews and informal conversations, conducted in a variety of locations and amid diverse activities, with frequent interruptions and digressions to match local communication norms. Communication behavior and content were grassroots, interactive, multitasking, and chaotic.

The author lived in Lasama for three months of ethnographic field research and made day trips into neighboring coastal villages and islands in the Tiworo Straits; this allowed her to observe and map social relations and communication networks and to listen to communication content seemingly unrelated to the fish pond, but which ultimately helped reveal the social, spatial, and temporal challenges that led to implementation failure. For WNP, the author made two three-week research trips in 2011 and 2013. She explored different settlements and spent a week living in Sama Bahari at the home of a community activist to observe communication and livelihood practices and another week staying at the offices of a local NGO in the nearby town of Ambeua to observe the community. She interviewed fisherfolk, local NGO members, village chiefs, and Wakatobi government officials, and engaged in casual conversations with residents and international ENGO staff. The aim was to assess communication and the effectiveness of public consultations regarding multiple-use zones.

There was rarely a one-on-one interview. As the chief breadwinner, the man of the household is the proper person to approach for an interview. Women typically declined interviews and volunteered their spouse. Interviews generally started with one person, and family, friends, neighbors, even children would join in and comment uninvited, with the result that interviews devolved into multidirectional conversations.

Communication Time-Space

Communication in the two communities was socially, spatially, and temporally distant from deliberative arenas. Informal communication flows are part of everyday lived experience. Deliberative participation did not suit the local communication environments.

Homes here are typical of fishing communities across Indonesia, constructed of mangrove wood on stilts above the ground or water to catch the cooling breeze. On land, the space below houses can be used as a secondary sleeping area, a place to hang laundry, an outdoor shelter from the sun to perform chores such as repairing fishing nets, preparing firewood, carving souvenirs, and washing. On water, the space below houses allows the sea to remove human waste and other garbage dumped directly from above; it is also a space to moor boats. Children play below houses on both land and sea. Because these are outdoor spaces, anyone can walk by to observe and talk. They are private spaces open to public view.

Designated public spaces include the simply built village hall and mosque. Both villages have a common outdoor recreational area, a field by a side road in Lasama, and in Sama Bahari, a dirt area on a coral atoll, surrounded by houses, with nets strung around the perimeter to prevent the balls landing in the sea. Lasama is on solid ground with wide spaces between houses; its main road serves as a busy thoroughfare for vehicular traffic en route to its port, the main gateway to the heavily populated islands in the Tiworo Straits. Sama-Bahari sits on a coral atoll that can only be reached by boat; there is no vehicular traffic on the atoll, and footbridges about two meters wide connect homes and public spaces throughout the village. These spaces are busy gathering places for residents in late afternoon and well into evening.

Daily activities are typical of maritime life in these parts. Fishermen usually leave at night for their fishing grounds and return in the early morning to go home and sleep. The hours from noon until about 3:00 p.m. are the hottest parts of the day, when most villagers stay close to home in the shade. This fishing cycle runs every two days with the third day off to rest and repair gear, although it varies according to individual preference and earnings aspirations. Fisherwomen typically do not go to sea at night but focus on near-shore or low-tide extractive activities such as the gathering of mollusks and abalone; women in Wakatobi gather seaweed and, before the practice was banned when WNP was established, mined coral from the reef for sale as building material. Some women also engage in small-scale goods and services, waking before dawn to prepare foodstuffs for sale in the village, as well as setting about their daily household chores.

Village life is quiet until the sun wanes. After 3:00 p.m., villagers are more likely to wander outdoors, visit with neighbors, gather in public spaces, socialize, and catch up on news. Long conversations are held outdoors among villagers standing, sitting, or squatting along the thoroughfares. This is the time to observe the public agora as a democratic space to debate issues of community interest, a marketplace of exchange and engagement, and a space where different publics—men, women, children, fishers, traders, leaders—gather in collective communicative action. These thoroughfares and publics constitute habitual, informal agoras, yet they are easily bypassed by outside managers, researchers, and NGOs intent on *traveling through* these spaces and publics to the deliberative destinations where they will engage with invited participants. These maritime agoras form a missing piece in participation.

COMMUNICATION AND MANAGEMENT FAILURE

Communication and management failures occurred amid interventions that—in research interviews, project reports, and online gray literature— government authorities and outside NGOs portray as community-based and participatory resource management. This was disputed by villagers and local NGO volunteers during interviews and social gatherings.

In Lasama, a 10-hectare stream fish pond was built on the edge of mangrove forest, with gates along the clear-cut perimeter trench to trap fish that had entered the pond at high tide but would be contained within at low tide. The design required only 20% per cent of the pond to be clear-cut, preserving mangroves and making use of naturally occurring saltwater and freshwater systems, an environmentally friendly design compared to the intensive cultivation model of inland clear-cut saltwater shrimp aquaculture. The Kendari-based NGO that designed the pond project invited select families to information meetings after consulting the village chief, with the result that 25 of the most socially connected and affluent households were informed about the project and became part of the collective. By contrast, the project was designed to support 10 of the poorest families in the village. The NGO named Women in Unity ostensibly promoted women's interests, but no women were in the project. The NGO staffer never visited the area again because of illness. A colleague with another NGO on Muna stepped in, but by then the members of the collective had stopped cooperating and had run out of money.

In WNP, the original multiple-use zoning system in 1996 was criticized as detrimental to local livelihoods, with poor participation and unclear communication. As a result, a public consultation called the rezonasi was completed in 2007 with the intent of redesigning the zones through more responsive strategies based on community input. But research revealed villagers were approached mainly to extract information about their knowledge about species distribution and health, and fishing practices such as location and equipment. Research respondents said they had no input on actual decisions and little opportunity to voice their livelihood priorities. Many remain confused about prohibited zones. Apparently a copy of the zoning map, on letter-sized paper, was distributed to the central government office of each of the four main islands in Wakatobi, but no village respondents were aware of it, aside from staff at community NGOs.

Public consultations consisted of meetings and workshops. Village chiefs were asked to disseminate word of upcoming meetings to key local leaders and livelihood interest groups. In Lasama, the NGO involved held a single information meeting with villagers to "sell" a predesigned fish pond project. In Sama Bahari, villagers were expected to answer questionnaires designed by the Nature Conservancy in collaboration with WWF written in technical and scientific language. It was then delivered to the village chief to forward to villagers, many of whom are barely literate with minimal schooling. The village chief warned TNC-WWF staff that this communication method would not work but was not heeded. TNC-WWF contracted local facilitators, who held one meeting in each targeted village toward the end of scheduled consultations to explain the new zoning design. Through the activities described above, participation was conducted according to deliberative norms that reinforced formal exercises with select participants.

Good participation is premised on good communication. However, in Lasama and Sama Bahari, it is doubtful whether good communication occurred. Indeed, the two sites are distinguished by communication and management failure. By villagers' reckoning, Lasama's fish pond and WNP failed to deliver to the poorest people. Their livelihoods have not improved; the fishery continues to yield less catch for more effort, though other Wakatobi villages are faring better. When questioned on the quality and frequency of communication with outside authorities and NGOs, villagers' responses ranged from "never" to "every month" to laughter. In these communities, Dyer et al.'s (2014) criteria for management failure came true: poor quality of project design, unqualified actors entrusted with implementation, and low transfer of power and authority to marginalized actors. Communication was infrequent, unevenly disseminated to relevant parties, poorly understood, and ultimately led to distrust of government, outside NGOs, and quite likely researchers. Direct communication with villagers was minimal and more top down than bottom up, reflecting a failure of participatory resource management.

Conclusion: The Maritime Agora as Communicative Space

The agora implies that space is a dimension of communication with a physical and social reality. Marine spaces pose communication challenges by nature of their patterns of dispersion, infrastructure development, cultural communities, and social norms. They are fluid and boundless, supporting diverse aquatic species and human populations, local livelihoods, national economic prerogatives and global demand for marine products, goods shipment, and, increasingly, conservation. Tiworo and WNP show that resource management interventions risk failure if public participation is not attuned to local communication environments. To engage communities more effectively, the design of participation needs to accommodate both deliberative arenas and the informal spaces of everyday communication. The maritime agora describes a communicative space that is chaotic, interstitial, and hybrid. Recognition of this agora expands the scope of public participation design-practice-assessment by exploring how current participation structures interact with spatially specific communication practice, environments, and authorial voices. Effective community engagement requires the design of spatially informed communication practice that disrupts deliberative participation infrastructures that may reproduce communication bottlenecks and thus open the potential to reassemble participation structures and practice befitting local spatial realities and social norms.

Communication behavior in the maritime agora can seem chaotic and disorderly, but that is intrinsic to informal participation. People invite themselves into interviews, they interrupt, and they do chores while being interviewed. This is the opposite of communication behavior at an orderly public consultation. However, chaotic communication can produce unsought information that reveals social structures and other challenges that compromise a resource management strategy, such as gender exclusion from certain activities. Chaotic communication signals that communication methods need to be creative to access people's stories and lived experiences.

The lessons of Lasama and WNP invite a reimagining of the location of communication as interstitial or *between* space. When outsiders traversed the village to reach meetings, they bypassed the spaces of social life. Deliberative spaces of communication serve literally and metaphorically as designated destinations *to which people travel* with the aim of discussing specific topics. However, private spaces and thoroughfares are not just temporary passing spaces; they are habitual time-spaces of engagement. These interstitial spaces lie between and beyond the arenas of invited participation, yet they are critical spaces for knowledge sharing.

Space has shifting meanings. Spatial hybridity means that the Indonesian maritime agora is a fluid space where private life is performed in public view, and private space becomes the temporary stage for public discourse. Where deliberative participation space is open only to a select few, excluded publics have recourse to habituated zones of engagement where their private selves engage publically in social life. In Tiworo and WNP, these public-private spheres are communicative spaces framed by roads, footbridges, water wells, and backroom workspaces.

Tiworo and Wakatobi are connected by similar histories of marine resource dependence. Their inhabitants share stories in spaces that predate new spatial imaginaries or participatory decision-making processes. Their stories, their norms of communication, and their zones of engagement need to be recognized as valid sites of knowledge. These two fisheries present cautionary tales of resource management and communication failure, but also overlooked spaces and opportunities for participation. The maritime agora of informal, nondeliberative everyday interactions is key to moving toward a communicative space for marine resource managers, scientists, conservation interests, and local communities in both Indonesia and beyond.

The maritime agora is a space of possibility that shifts with the social, spatial, and temporal context, but is always a space where publics exchange information, aspirations, and challenges as they perform their relationships within their communities. Recognizing the existence of such spaces, and accounting for how, what, and which publics communicate in such spaces, is critical to good participation. These insights are particularly relevant in developing countries where logistical and infrastructure challenges of space can compromise effective participation—this insight applies to both marine and terrestrial systems. The spatiality of communication is critical to understanding public participation that transcends particular locales, environments, and marginal publics.

Ultimately, we need to treat environmental communication as a culturally specific phenomenon; the context for communication is not universal. Communication activity occurs beyond the participation spaces that are conferred validity by formal authorities or their proxies such as NGOs. Any type of participation or communication space should be culturally grounded and responsive to the particular public sphere affected by resource policy decisions.

References

Abah, O. S. (2007). Vignettes of communities in action: An exploration of participatory methodologies in promoting community development in Nigeria. *Community Development Journal, 42*(4), 435–448. doi:10.1093/cdj/bsm035

Adams, P. C., & Jansson, A. (2012). Communication geography: A bridge between disciplines. *Communication Theory, 22*(3), 299–318. doi:10.1111/j.1468-2885.2012.01406.x

Bennett, N. J., & Dearden, P. (2014). Why local people do not support conservation: Community perceptions of marine protected area livelihood impacts, governance and management in Thailand. *Marine Policy, 44*, 107–116. doi:10.1016/j.marpol.2013.08.017

Bowlby, S. (2012). Recognising the time-space dimensions of care: Caringscapes and carescapes. *Environment and Planning A, 44*(9), 2101–2118. doi:10.1068/a44492

Bryant, A. (2006). *Wiki* and the *Agora*: "It's organising, Jim, but not as we know it." *Development in Practice, 16*(6), 559–569. doi:10.1080/0961520600958165

Castells, M. (2008). The new public sphere: Global civil society, communication networks, and global governance. *Annals of the American Academy of Political and Social Science, 616*, 78–93. doi:10.2307/25097995

Chambers, R. (1997). Editorial: Responsible well-being—a personal agenda for development. *World Development, 25*(11), 1743–1754. doi:10.1016/S0305-750X(97)10001-8

Chambers, R. (2008). *Revolutions in development inquiry*. London: Earthscan.

Chuenpagdee, R., Pascual-Fernández, J. J., Szeliánszky, E., Luis Alegret, J., Fraga, J., & Jentoft, S. (2013). Marine protected areas: Re-thinking their inception. *Marine Policy, 39*(1), 234–240. doi:10.1016/j.marpol.2012.10.016

Coleman, E. A., & Mwangi, E. (2013). Women's participation in forest management: A cross-country analysis. *Global Environmental Change, 23*(1), 193–205. doi:10.1016/j.gloenvcha.2012.10.005

Cons, J. (2016). Conclusion: The placial imagination. *Journal of Environmental Studies and Sciences*, *6*(4), 788–789. doi:10.1007/s13412-015-0262-8

Cooke, B., & Kothari, U. (Eds.). (2001). *Participation: The new tyranny?* London and New York: Zed Books.

Cornwall, A., & Coelho, V. S. P. (Eds.). (2007). *Spaces for change: The politics of citizen participation in new democratic arenas*. London and New York: Zed Books.

Cornwall, A., & Pratt, G. (2011). The use and abuse of participatory rural appraisal: Reflections from practice. *Agriculture and Human Values*, *28*(2), 263–272. doi:10.1007/s10460-010-9262-1

Cox, R. (2013). *Environmental communication and the public sphere* (3rd ed.). Thousand Oaks, CA: Sage Publications.

Davenport, S., & Leitch, S. (2005). Agoras, ancient and modern, and a framework for science–society debate. *Science and Public Policy*, *32*(2), 137–153. doi:10.3152/147154305781779605

de Nooy, W., & Nooy, W. De. (2013). Communication in natural resource management: Agreement between and disagreement within stakeholder groups. *Ecology and Society*, *18*(2), art44. doi:10.5751/ES-05648-180244

de Souza, H. N., Cardoso, I. M., de Sá Mendonça, E., Carvalho, A. F., de Oliveira, G. B., Gjorup, D. F., & Bonfim, V. R. (2012). Learning by doing: A participatory methodology for systematization of experiments with agroforestry systems, with an example of its application. *Agroforestry Systems*, *85*(2), 247–262. doi:10.1007/s10457-012-9498-4

de Souza, R. (2009). Creating "communicative spaces": A case of NGO community organizing for HIV/AIDS prevention. *Health Communication*, *24*(8), 692–702. doi:10.1080/10410230903264006

Diehm, C. (2012). Biophilia and biodiversity: Environmental ethics in the work of Stephen R. Kellert. *Environmental Ethics*, *34*(1), 51–66.

Dyer, J., Stringer, L. C., Dougill, A. J., Leventon, J., Nshimbi, M., Chama, F., . . . Syampungani, S. (2014). Assessing participatory practices in community-based natural resource management: Experiences in community engagement from southern Africa. *Journal of Environmental Management*, *137*, 137–145. doi:10.1016/j.jenvman.2013.11.057

Elliot, G., Mitchell, B., Wiltshire, B., Abdul Manan, I. R., & Wismer, S. (2001). Community participation in marine protected area management: Wakatobi National Park, Sulawesi, Indonesia,. *Coastal Management*, *29*(4), 295–316. doi:10.1080/089207501750475118

Fischler, R. (2000). Communicative planning theory: A Foucauldian assessment. *Journal of Planning Education and Research*, *19*(4), 358–368. doi:10.1177/0739456X0001900405

Flyvbjerg, B. (1998). Habermas and Foucault: Thinkers for civil society? *The British Journal of Sociology*, *49*(2), 210. doi:10.2307/591310

Flyvbjerg, B., & Richardson, T. (2002). Planning and Foucault: In search of the dark side of planning theory. In P. Allmendinger & M. Tewdwr-Jones (Eds.), *Planning Futures: New directions for planning theory* (pp. 44–62). London and New York: Routledge. doi:10.1177/1473095208090432

Fortmann, L. (Ed.). (2008). *Participatory research in conservation and rural livelihoods: Doing science together.* Hoboken, NJ: Wiley-Blackwell.

Freire, P. (1970). *Pedagogy of the oppressed.* New York, NY: Herder and Herder.

Garcia-Blanco, I., Van Bauwel, S., & Cammaerts, B. (Eds.). (2009). *Media agoras: Democracy, diversity, and communication.* Newcastle upon Tyne, UK: Cambridge Scholars Publishing.

Geoghegan, M., & Powell, F. (2009). Community development and the contested politics of the late modern agora: Of, alongside or against neoliberalism? *Community Development Journal, 44*(4), 430–447. doi:10.1093/cdj/bsn020

Glaser, M., Baitoningsih, W., Ferse, S. C. A., Neil, M., & Deswandi, R. (2010). Whose sustainability? Top-down participation and emergent rules in marine protected area management in Indonesia. *Marine Policy, 34*(6), 1215–1225. doi:10.1016/j.marpol.2010.04.006

Glucker, A. N., Driessen, P. P. J. J., Kolhoff, A., & Runhaar, H. A. C. C. (2013). Public participation in environmental impact assessment: Why, who and how? *Environmental Impact Assessment Review, 43*, 104–111. doi:10.1016/j.eiar.2013.06.003

Gomez, C. A. T. (2011). Communicative citizenship, preliminary approaches. *Signo Y Pensamiento, 30*, 106–128.

Habermas, J. (1984). *The theory of communicative action, Volume 1, Reason and the rationalization of society.* Boston: MIT Press.

Habermas, J. (1987). *The theory of communicative action, Volume 2, Lifeworld and system: A critique of Functionalist reason* (Vol. 2). Boston: MIT Press.

Hickey, S., & Mohan, G. (2004). Towards participation as transformation: Critical themes and challenges. In S. Hickey & G. Mohan (Eds.), *Participation: From tyranny to transformation?* (pp. 3–24). London and New York: Zed Books. doi:10.1007/s13398-014-0173-7.2

Hurlbert, M., & Gupta, J. (2015). The split ladder of participation: A diagnostic, strategic, and evaluation tool to assess when participation is necessary. *Environmental Science & Policy, 50*, 100–113. doi:10.1016/j.envsci.2015.01.011

Huxley, M. (2013). Historicizing planning, problematizing participation. *International Journal of Urban and Regional Research, 37*(5), 1527–1541. doi:10.1111/1468-2427.12045

Innis, H. A. (1986). *Empire and communications.* Victoria, BC: Press Porcepic Ltd.

Institute of Development Studies. (2016). Participation. Retrieved from http://www.ids.ac.uk/team/participation

Iyall Smith, K. E. (2007). New agoras and old institutions: The case of human rights. *Systemic Practice and Action Research, 20*(5), 387–399. doi:10.1007/s11213-007-9074-4

Janelle, D. G. (1968). Central place development in a time-space framework. *Professional Geographer, 20*(1), 5–10. doi:10.1111/j.0033-0124.1968.00005.x

Jenlink, P. M., & Banathy, B. H. (2002). The Agora Project: The new agoras of the twenty-first century. *Systems Research and Behavioral Science, 19*(5), 469–483. doi:10.1002/sres.502

Kajubi, P., Bagger, S., Katahoire, A. R., Kyaddondo, D., & Whyte, S. R. (2014). Spaces for talking: Communication patterns of children on antiretroviral therapy in Uganda. *Children and Youth Services Review, 45*, 38–46. doi:10.1016/j.childyouth.2014.03.036

Kallianos, Y. (2013). Agency of the street. *City, 17*(4), 548–557. doi:10.1080/136 04813.2013.812368

Kellert, S. R. (1996). *The value of life, biological diversity and human society.* Washington, DC: Island Press/Shearwater Books.

Ku, L., & Zaroff, C. (2014). How far is your money from your mouth? The effects of intrinsic relative to extrinsic values on willingness to pay and protect the environment. *Journal of Environmental Psychology, 40*, 472–483. doi:10.1016/j.jenvp.2014.10.008

Lee, C. W. (2007). Is there a place for private conversation in public dialogue? Comparing stakeholder assessments of informal communication in collaborative regional planning. *American Journal of Sociology, 113*(1), 41–96. doi:10.1086/517898

Levy, S. S. (2003). The Biophilia Hypothesis and anthropocentric environmentalism. *Environmental Ethics, 1*(1979), 227–246.

Lloyd, D., Van Nimwegen, P., & Boyd, W. E. (2005). Letting indigenous people talk about their country: A case study of cross-cultural (mis)communication in an environmental management planning process. *Geographical Research, 43*(4), 406–416. doi:10.1111/j.1745-5871.2005.00343.x

Marques, A. S., Ramos, T. B., Caeiro, S., & Costa, M. H. (2013). Adaptive-participative sustainability indicators in marine protected areas: Design and communication. *Ocean & Coastal Management, 72*, 36–45. doi:10.1016/j.ocecoaman.2011.07.007

Mitchell, B. (Ed.). (2015). *Resource and environmental management in Canada* (5th ed.). Toronto, ON: Oxford University Press.

Neumann, B., Vafeidis, A. T., Zimmermann, J., & Nicholls, R. J. (2015). Future coastal population growth and exposure to sea-level rise and coastal flooding—a global assessment. *PLoS ONE, 10*(3). doi:10.1371/journal.pone.0118571

Oyanedel, R., Marín, A., Castilla, J. C., & Gelcich, S. (2016). Establishing marine protected areas through bottom-up processes: insights from two contrasting initiatives in Chile. *Aquatic Conservation: Marine and Freshwater Ecosystems, 26*(1), 184–195. doi:10.1002/aqc.2546

Plot, M. (2009). Communicative action's democratic deficit: A critique of Habermas's contribution to democratic theory. *International Journal of Communication, 3*, 825–852.

Powell, F. (2012). "Think globally, act locally": Sustainable communities, modernity and development. *GeoJournal, 77*(2), 141–152. doi:10.1007/s10708-009-9330-5

Pullin, A. S., Bangpan, M., Dalrymple, S., Dickson, K., Haddaway, N. R., Healey, J. R., . . . Oliver, S. (2013). Human well-being impacts of terrestrial protected areas. *Environmental Evidence, 2*(1), 19. doi:10.1186/2047-2382-2-19

Reed, M. S. (2008). Stakeholder participation for environmental management: A literature review. *Biological Conservation, 141*(10), 2417–2431. doi:10.1016/j.biocon.2008.07.014

Rienstra, B., & Hook, D. (2006). Weakening Habermas: The undoing of communicative rationality. *Politikon, 33*(3), 313–339. doi:10.1080/02589340601122950

Rogers, R. A., & Kincaid, D. L. (1981). *Communication networks: Toward a new paradigm for research*. New York, NY: The Free Press.

Sager, T. (2013). *Reviving critical planning theory: Dealing with pressure, neo-liberalism, and responsibility in communicative planning*. New York, NY: Routledge (in conjunction with the Royal Town Planning Institute).

Tam, C.-L. (2006). Harmony hurts: Participation and silent conflict at an Indonesian fish pond. *Environmental Management, 38*(1), 1–15. doi:10.1007/s00267-004-8851-4

Tam, C.-L. (2015). Timing exclusion and communicating time: A spatial analysis of participation failure in an Indonesian MPA. *Marine Policy, 54*, 122–129. doi:10.1016/j.marpol.2015.01.001

Torrens, P. M. (2008). Wi-Fi geographies. *Annals of the Association of American Geographers, 98*(1), 59–84. doi:10.1080/00045600701734133

Touaf, L., & Boutkhil, S. (Eds.). (2008). *The world as a global Agora: Critical perspectives on public space*. Newcastle upon Tyne, UK: Cambridge Scholars Publishing.

Tremblay, G. (2012). From Marshall McLuhan to Harold Innis, or from the global village to the world empire. *Canadian Journal of Communication, 37*, 561–575.

Turnhout, E., Van Bommel, S., & Aarts, N. (2010). How participation creates citizens: Participatory governance as performative practice. *Ecology and Society, 15*(4), 26 [online].

van Zyl, M. (2009). Ocean, time and value: Speaking about the sea in Kassiesbaai. *Anthropology Southern Africa, 32*(1&2), 48–58.

Yates, K. L. (2014). View from the wheelhouse: Perceptions on marine management from the fishing community and suggestions for improvement. *Marine Policy, 48*, 39–50. doi:10.1016/j.marpol.2014.03.002

Chapter 9

Rare's Conservation Campaigns

Community Decision Making and Public Participation in Global Contexts

Sarah D. Upton, Carlos A. Tarin, Stacey K. Sowards,
and Kenneth C. C. Yang

Rare is an international organization aimed at achieving conservation results by partnering with other environmental organizations to train local leaders to develop and manage Pride campaigns. For more than 25 years, Rare has worked with local leaders, known as campaign managers (CMs), to design and implement Pride campaigns in more than 50 countries. These Pride campaigns use social marketing to empower local communities to take direct action in response to environmentally problematic behaviors such as illegal logging, overfishing, and misuse of coastal no-take zones. Because the proposed behavioral change originates from local leaders, communities are able to cultivate a "sense of ownership, responsibility and pride in conservation" (Rare, 2016a). Rare's approach involves identifying human behaviors that cause threats to biodiversity, using social science research to identify public participation solutions to change these behaviors, and launching a Pride campaign, designed to instill pride within a local community and to facilitate the removal of participatory barriers to conservation and the adaption of conservation solutions on a broader scale. Each Pride campaign follows a formula known as Rare's Theory of Change (TOC), which is culturally adapted to be suitable for different communities, ranging from rural to more urban environments. This formula is customized to meet the unique needs of each Pride campaign while following general guidelines for success based on "the factors, conditions, and sequence of events that led to positive change" (Rare, n.d.–b, p. 8). The components include Knowledge + Attitude + Interpersonal Communication + Barrier Removal => Behavior Change => Threat Reduction => Conservation Result. The idea behind this model is that the first three factors of knowledge, attitude, and interpersonal communication are necessary persuasive elements needed to remove a barrier in a community that is preventing conservation effectiveness. Once a barrier has been removed or mitigated, then behavior change and threat reduction can occur, which leads to the desired conservation result.

In this chapter, we explore the ways in which Rare uses knowledge as a symbolic resource to enable stakeholders' behavioral changes, which subsequently leads to environmental conservation. Rare partners with the University of Texas at El Paso to offer a master's degree program as part of its approach to implement behavior change in conservation-related activities. Each of the authors of this chapter has coordinated one of the language programs as part of our master's degree that is offered in Indonesia in Indonesian, China in Mandarin Chinese, and Latin America in Spanish. Methodologically, each of us has collected data while supervising the coursework and assignments for this program through qualitative approaches, such as ethnography, interviews, and field site visits; and quantitative approaches, such as knowledge-attitude-practice (KAP) surveys implemented by our students (the Rare campaign managers or CMs). We begin with an overview of change paradigms and a description of the three levels of organizational knowledge operating in Rare's approach to conservation, which emphasizes public participation. We then offer examples from three of the regions where Rare works: Indonesia, Latin America, and China. We ultimately argue that Rare's focus on symbolic resources approaches conservation through the paradigm of *constructed potentiality*, generating multiple options for community-driven decision making and public participation at international, national, regional, and local levels. Such an approach enables Rare and its campaign managers to draw on expertise from all kinds of backgrounds, experiences, and different knowledges and ultimately allows for contextual and effective behavioral change in conservation rooted in public participation.

Constricted and Constructed Potentiality in Public Participation

Modern threats to ecosystems and biodiversity compel activists, scholars, practitioners, and politicians alike to advocate for environmental change through public participation. To understand how Rare's approach to creating change through Pride campaigns differs from other types of conservation efforts and other approaches to public participation, we explore the change paradigms of *constricted potentiality* and *constructed potentiality.* Foss and Foss (2011) explain that actors go through a series of steps when seeking to create change, beginning with an initial choice to focus on either material conditions or symbols as resources. This initial choice determines the strategy for achieving change, the route to change, the focus of change efforts, and ultimately the outcomes that will result. Dominant theories of change, or constricted potentiality, limit the available options for creating change by focusing on tangible material conditions. This focus on tangible resources means that they are "fixed, limited, and capable of being depleted" (Foss & Foss, 2011, p. 208). A focus on material resources means that persuasion becomes the strategy for change, and change agents develop a prescribed route to change. The focus of change efforts is external, and the result is a change in material conditions.

Similarly, contemporary approaches to public participation provide a wealth of theoretical concepts such as standing, influence, and access (Senecah, 2004); collaboration and trust (Daniels & Walker, 2001); and methods for addressing the participation gap (Callister, 2013; Cox, 2004), which are all intrinsic to successful public participation practices. However, many of these approaches do not fully theorize the role of knowledge formation, acquisition, or use in engaging stakeholders and, in fact, often treat knowledge as synonymous with expertise or scientific fact. In many of these approaches, knowledge is used for purely instrumental means to persuade publics to engage in particular ways.

In addition to not always producing desired results, change efforts based in persuasion are often rooted in a desire to control others, and this desire to persuade and control means a devaluing of the life-worlds of others (Foss & Foss, 2012; Foss & Griffin, 1995). This approach to change also limits opportunities for public participation, as persuasion is about conquest and conversion (Foss & Foss, 2012; Ryan & Natalle, 2001), which makes it a colonizing force. This means that when there is an attempt to change or control another person, it is grounded in ideas of entitlement and superiority; the persuader assumes they have the right to change the other person in the interaction. This approach results in colonizing the life-worlds of others, much like the colonization of indigenous territories and ways of life perpetuated by the West. Gorsevski (2004) argues that this desire to control others is a violent form of rhetoric, as it can be viewed as an attack on individuals' ways of life.

In response to the paradigm of constricted potentiality, which uses persuasion as a change strategy, Foss and Foss (2011) offer an alternative change paradigm, constructed potentiality, in which the public's initial choice is to focus on the symbols available for creating change. The focus on symbolic resources means that unlimited resources become available because individuals "never run out of new ways to configure and construct symbols" (Foss & Foss, 2011, p. 213). This, in turn, creates more opportunities for public participation, as a greater number of people have access to resources to create change. A focus on symbolic resources leads to interpretation as the strategy for change and an unspecified route to change. The focus of change efforts is internal, and the result is self-change.

We argue that Rare's approach to conservation through the creation of Pride campaigns illustrates the importance of the paradigm of constructed potentiality and more complex conceptualizations of knowledge as a tool for participatory engagement. Pride campaigns focus on knowledge as a symbolic resource that can be identified, cultivated, and shared to create change. Campaign managers use interpretation as a strategy for change, choosing to view local knowledge as an abundant resource in their campaigns. The focus of Rare's change efforts is internal, and the outcome of efforts is self-change (or at least within the community), as both campaign managers and community members choose how they use knowledge to evaluate and change practices in their own lives to contribute to local conservation efforts.

Our analysis expands on existing public participation literature by situating knowledge as a complex, relational accomplishment that can be used as a valuable resource in the tool kit of both academics and practitioners. In focusing on knowledge as a form of constructed potentiality, we aim to highlight how such participatory approaches can serve as a valuable heuristic that avoids the potential pitfalls of frameworks focused primarily on institution-level change (Burns & Uberhost, 1988; Waddell, 1996). Constructed potentiality exemplifies what Gorsevski (2004) calls "nonviolent rhetoric," which is creative, positive (focusing on the strengths of social movement participants), and sustainable; and which does not vilify the opposing side of the issue. Nonviolent rhetoric is characterized by respect for and awareness of culture, a reliance on community and mutual responsibility, a refusal to engage with unjust actions and systems, and an underdog ethos (Gorsevski, 2004). To this end, we conceptualize Rare's style of public participation as an exercise in defining, creating, and sustaining particular forms of knowledge that are used in conservation campaigns to achieve constructed potentialities that empower substantive social change. In the following sections, we conceptualize three levels of organizational knowledge in Rare's approach to conservation and provide regional examples from Indonesia, Latin America, and China to further illustrate the ways in which knowledge is used as a symbolic resource through constructed potentiality in Pride campaigns.

Rare and Organizational Knowledge

Rare's unique approach to public participation relies on a complex coordination of various types of knowledge to facilitate long-term engagement and capacity building. Unlike institutional models of knowledge transmission wherein communities are passive recipients of scientifically proven "best practices," Rare's model of public participation encourages community members to critically engage with the conditions endemic to their environmental context. In taking such a localized approach to organizing, Rare aims to improve environmental outcomes while simultaneously building capacity of campaign managers, local organizations, and communities. This is noteworthy because, as Lindenfeld et al. (2012) explain, "One of the greatest challenges facing sustainability science results from difficulties in producing knowledge to support improved decision making, while strengthening communities' socioeconomic and cultural assets. Traditional methods of generating knowledge are ill equipped to further this broader goal" (p. 27). Rare's approach to community organizing relies on a nontraditional method of knowledge generation in which important goals, strategies, and outcomes are co-constitutively developed by a complex interplay of various levels of knowledge claims. Thus, Rare provides an opportunity to understand both the practical and theoretic dimensions of organizational knowledge as a symbolic resource that can be employed in public participation contexts.

Organizational knowledge is an area of scholarship that has existed since the early days of organizational theorizing, although the term has changed according to the circumstances of its use and application (Canary & McPhee, 2011; Hodgkinson & Healey, 2008; Lam, 2000). Early organizational studies, dominated by social psychological perspectives, tended to understand knowledge as a type of currency that could be used in employee-employer relations to maximize productivity (Perrow, 1986). As Canary and McPhee (2011) explain, this trend (Taylorism) assumed that knowledge was something that could tangibly be assessed and measured. Since that time, organizational scholars have broadened their conceptualization of knowledge processes and the complex ways in which knowledge is constituted, maintained, and transmitted to members of an organization (Barge & Schockley-Zalabak, 2008; Dulipovici & Baskerville, 2007; Kuhn & Porter, 2011; Leonardi, 2011).

We understand knowledge to be a relational accomplishment that includes both individual and collective components. McPhee, Corman, and Dooley (1999) note that organizational knowledge is "the symbolic and/ or practical routines, resources, and affordances drawn on by organization members and social units as they maintain the institutional organization and/ or coordinate their action and interaction" (p. 4). Organizational knowledge, then, can be understood as inherently communicative in nature and linked to a variety of issues such as organizational culture and organizational memory. Our analysis in this chapter is predicated on two prevailing assumptions about organizational knowledge. First, we understand knowledge to be a frame for structuring communicative practices and organizational reality (Lakoff, 2010) that can be used to bolster public participation efforts. By frame, we mean that knowledge serves as a sort of filter that shapes and influences discourses in particular ways so as to create shared meanings among members of the organization. Because frames are not static, knowledge becomes something that is permeable and sometimes contested. Scholars such as Barley, Leonardi, and Bailey (2012) and Murphy and Eisenberg (2012) have argued that knowledge can be used productively as a resource for creating both clarity and ambiguity in ways that may be negotiated and politicized by members of the organization. Thus, our analysis is guided by the assumption that knowledge serves as a discursive frame of reference that can be apprehended by members of an organization to accomplish partici-patory outcomes. Second, we understand knowledge to be process oriented and something that can be done interactively. As Canary (2010) asserts, "Developing organizational knowledge does not occur as a singular event; rather, the process occurs iteratively during many different organizational activities that involve different goals" (p. 27). In contending that knowledge is something that is done or accomplished, we mean to suggest that knowledge is constitutively formed and, more importantly, maintained by adherence to particular forms of intelligibility in the organization. Knowledge, though collective, must be (re)performed in practice to cohere to prevailing standards and assumptions. In this sense, organizational knowledge can be understood through repeated acts and sustained discourses. Approaching knowledge as both a frame and interactive process reinforces its potential as a symbolic resource by positioning knowledge as something that individuals and entire communities can draw on and perform in an effort to create environmental change through constructed potentiality.

Within the context of public participation, knowledge serves as a structuring element that is used to facilitate and coordinate meanings and actions within organizations and communities. That is, knowledge must be learned by members of the organization before it can become usefully employed in public practice. These messages are transmitted both within and across systems of activity. As Canary (2010) explains, "[By] focusing on communication within and between activity systems [we] might extend not only the model of activity systems, but . . . also change current conceptualizations of the process of constructing knowledge" (p. 36). This study, then, aims to explore the practices and discourses that are used to constitutively construct knowledge as an unlimited symbolic resource within Rare and the ways in which these knowledge creation and acquisition processes shape meanings and activity within the organization to more readily instigate public participation through constructed potentiality.

To this end, we conceptualize three distinct levels of organizational knowledge that serve as a schema for framing how Rare's campaign managers use, coordinate, and build on knowledge processes in practice. These levels—which we term institutional, practical, and local—enable Rare's campaign managers to use knowledge as a symbolic resource to construct change potentiality through their activities in the field. Institutional knowledge refers to organizational discourses and meanings that are standardized throughout Rare's approach. Practical knowledge, by contrast, refers to knowledge that is acquired through practice and experience. As a departure from institutional knowledge, practical knowledge deals with the type of information that may be passed on from member to member in non–officially sanctioned ways. Finally, local knowledge refers to the processes in which campaign managers draw on specific cultural, geographic, or sociopolitical understandings that are endemic to their region to create adaptive and responsive campaigns. This emphasis on locality is a key element of Rare's outreach strategy. Collectively, these levels of organizational knowledge enable members to more fully actualize participatory practices in ways that lead to more meaningful and sustained environmental change.

Organizational Knowledge in Rare Pride Campaigns

Institutional Knowledge

In this section, we highlight examples from Pride campaigns using institutional knowledge to increase public participation in Indonesia, China, and throughout Latin America. To achieve conservation results, Rare draws on institutional knowledge as a symbolic resource in various forms, including the Rare-UTEP partnership's curriculum, the TOC model, and barrier removal operation plans. Rare identifies conservation "bright spots," or problem areas where Pride campaigns can work to inspire conservation efforts in multiple communities, and selects specific campaign sites in areas with high levels of biodiversity and conservation challenges. Focusing on tropical and subtropical regions of the world, Rare works to engage people in creating and sustaining solutions to conservation challenges. For example, in Indonesia, Rare-UTEP cohorts have focused on the overarching theme of "Fish Forever," a global campaign aimed at protecting marine ecosystems and the region's extensive coral reef systems, while at the same time meeting the needs of local communities who depend on fish almost entirely for their food source and livelihoods (Rare Indonesia, 2015). In Guadalajara, Rare's central hub for Latin America, cohorts have focused on the use of reciprocal water agreements to protect watersheds. Based on the idea of reciprocity, these agreements "involve funding from downstream users that incentivizes [upstream] farmers to set aside part of their land for conservation" (Alger, 2014, p. 1). As of 2016, Rare has conducted 17 Pride conservation campaigns in China, ranging from endangered species protection, to wetland conservation, to sustainable agriculture (Rare, 2016b) to respond to the growing number of environmental problems in China, such as air and water pollution, desertification, and erosion (Yang, 2015).

After selecting conservation bright spots, Rare works to identify partners who will support the design, execution, and monitoring of the Pride campaigns, and each of these partners contributes to the construction of institutional knowledge as a symbolic resource. Implementing partner organizations provide local leaders, known as campaign managers, who enroll in Rare's training program and lead a Pride campaign. These campaign managers also become graduate students enrolled in the Master of Arts program in the department of communication at the University of Texas at El Paso. They complete their coursework in three university phases and three field phases over a two- to three-year period. Campaign managers begin by attending an approximately nine-week-long university session focused on behavior change, after which they return to their individual sites, obtain baseline survey data about barriers to change that exist in the community, and identify stakeholders who will participate in planning for the Pride campaign. CMs then return for a second five-week university phase to finalize their Pride campaign strategy, aimed at promoting public participation through cultivating environmental Pride and the adoption of sustainable alternative behaviors in their communities. Afterward, CMs then return to their sites, launch their Pride campaigns, and conduct community surveys to assess social and environmental change. With these data, fellows return for their third and final university phase to share their results and develop a plan for sustaining the impact they have made. After graduation, CMs continue to work with community members to apply techniques from their Pride campaigns to sustain and broaden their impact through the implementation of further campaign activities or the initiation of a new campaign in another (often nearby) community.

Rare also works with strategic, technical, and funding partners. Strategic partners, or experts working in conservation such as NGOs or government agencies, use their expertise to support the strategy, design, and sustainability of Pride campaigns. Technical partners provide technology or training, create policy, and/or conduct biological monitoring of Pride campaigns. Finally, funding partners not only provide funding for Pride campaigns, but also participate in adapting the TOC to meet the needs of each unique conservation bright spot. Although technical knowledge is functionally important, we conceptualize institutional knowledge as a more complex symbolic resource insofar as it involves the mediation of expertise with participatory engagement to accomplish tangible, targeted outcomes. Each stakeholder brings a combination of knowledge grounded in expertise, culture, lived experience, and/or theory, thus creating a network of unlimited, sustainable symbolic resources to draw from.

In China, Rare has used TOC, along with other social marketing tools, techniques, and theories, to protect endangered species (such as cranes, tigers, and monkeys) in different geographical locations in a vast country of 1.3 billion. Pride campaigns in China demonstrate how knowledge is generated, disseminated, and applied to develop grassroot campaigns to encourage people's pro-environmental behaviors. As stated in Rare's sustainable agricultural campaign in China that was recently launched in 2015, Rare's Pride campaigns are designed to train campaign managers "to inspire communities to adopt change through a Pride campaign, . . . to canvas their communities to determine what barriers to sustainable behaviors exist and create pathways to change, [and] . . . to collect information throughout their campaign and adapt their plans based on constant data analysis" (Rare, 2015). Campaign managers collected local residents' attitudes, beliefs, and interpersonal communication channels, all of which serve as symbolic resources available to create desirable changes through constructed potentiality.

Once the campaign managers have interpreted local residents' pre-campaign state of mind (as measured by their attitudes, beliefs, and knowledge), their interpretation of the circumstance is situated within the ecological system of the community to identify cultural, economic, political, social, and technological factors that affect the success of Pride campaigns. The removal of barriers to local residents' pro-environmental behaviors is critical to the success of Pride campaigns by employing approaches that provide non-mainstream and alternative solutions that are not supported or endorsed in the existing economic structure in China, again relying on symbolic rather than material resources. In the end, it is the ultimate objective of Pride campaigns to initiate local residents' behavioral change, leading to the reduction of threats and the protection of the environment. This behavioral change is an opportunity to channel change efforts internally, ultimately creating opportunities for self-change through constructed potentiality.

In Latin America, the TOC facilitates the process of obtaining reciprocal water agreements by using institutional knowledge as a symbolic resource to create awareness around the role ecosystems play in protecting clean water sources. To address attitudes surrounding upstream land use, campaign managers demonstrate the benefits of reciprocal water agreements and the social, cultural, and economic value of protecting their environment. Campaign managers use interpersonal communication to invite public participation through conversations about reciprocal water agreements among community stakeholders. To remove barriers, campaign managers introduce experts and technical support to help create and enforce these agreements. Behavioral change involves ensuring that conservation becomes a key factor in land management decisions for both farmers and municipal leaders. To demonstrate threat reduction and the conservation result, campaign managers monitor the reduction in damage to upstream ecosystems where reciprocal water agreements have been established, and ultimately demonstrate the protection of those ecosystems, reduction of endangerment to target species, and effective management of water resources. As these cohorts began their individual Pride campaigns aimed at watershed protection, Rare's institutional knowledge surrounding the use of the TOC to create reciprocal water agreements was an important symbolic resource that made it possible to achieve significant conservation results. In short, institutional knowledge functions as a guiding heuristic that orients Pride campaigns toward participatory outcomes.

Practical Knowledge

Pride campaigns also develop practical knowledge gained through practice and experience. Practical knowledge is an important symbolic resource used to increase public participation by identifying the unique characteristics of each conservation site and developing strategies for change that work for specific communities through constructed potentiality. Additionally, campaign managers work to identify effective communication channels that aid in increasing participation and create opportunities for community members to play an active role in campaigns through participatory practices such as community mapping and patrolling.

To successfully plan and implement Pride campaigns to address the unique characteristics of each location where Rare's institutional knowledge, such as how TOC is applied, campaign managers often rely on both qualitative and quantitative data collected on site to provide location-specific solutions. For example, in the Yang-Tze River Alligator Conservation Campaign in China, the campaign manager applied Rare's TOC framework to analyze many ecological factors affecting the conservation of endangered alligators in this area. The conservation of alligators was found to be contingent on whether or not community members were able to recognize the cultural importance of the protected species and if alternative solutions could be made available. However, when scrutinizing what facilitates the dissemination of conservation knowledge, the campaign manager realized the indispensable role of interpersonal communication channels in the rural Chinese community where non-official channels are more effective than mass media. An unanticipated text-to-mobile campaign was used to examine how local residents receive conservation information from community committees. A total of 1,256 mobile devices received the text. Response rates ranged from 4.4% (location #1) and 3.7% (location #2). The campaign manager analyzed and speculated that the cause of the low response rate was local residents' distrust of text messaging. In the Rare university phase, when all campaign managers returned to the Southwestern Forest University at Kunming to receive programmatic training of Rare's curriculum, this campaign manager shared his experience and approach of responding to an unanticipated circumstance on site. Through sharing his practical knowledge of handling this challenge, the campaign manager has added to institutional knowledge as a symbolic resource for future campaign managers.

At the level of practical knowledge, Rare's office staff members in Bogor, Indonesia, also work extensively with campaign managers to develop conservation strategies that have been known to work well in Indonesia. Rare's staff and the campaign managers chosen to participate in each cohort are all quite familiar with national-, regional-, and local-level problems, ranging from food security, poverty, health care, and educational access to more immediate problems such as natural disasters, all issues connected directly or indirectly to environmental threats. Standardizing barrier removal strategies is common, with a few key approaches that have been proven to work either in other campaigns or other Indonesian contexts. The primary environmental concern for these cohorts has been overfishing and marine ecosystem health. To prevent overfishing and restore ecosystems, Rare has extensively trained campaign managers about no-take zones and core zones (Rare Indonesia, 2015). A no-take zone is an area that has been determined to have biodiversity richness where a national, regional, provincial, local, or community law or agreement declares that no one will enter or use that zone. No-take zones or core zones are established and created through key stakeholder and community meetings in which local people determine where such zones should be located based on conservation principles and socioeconomic needs. That is, these zones, in extensive conversations with international and national biologists, conservationists, and local people, are based not just on protecting the ecosystem, but also on what will best serve the people. The participatory nature of creating no-take zones exemplifies an unspecified route to change in constructed potentiality, as multiple possibilities and options emerge through the focus on practical knowledge as a symbolic resource. Each campaign manager learns about how no-take zones work best and how to implement them through community engagement.

During the field phase of the campaigns, extensive key stakeholder and community member meetings are held to encourage widespread participation and generate community support and input for the no-take zone. By channeling change efforts internally, community members and local leaders decide where to establish the no-take zone and how to create awareness. For example, in campaigns in Karimunjawa National Park (implemented by Yusuf Syaifudin) and Kepulauan Seribu National Park (implemented by Yuniar Ardianti), community members developed and installed buoy markers so that everyone would know where the no-take zone was located (Rare Indonesia, 2014). Problematically in Indonesia, national parks are often created without effectively informing the local communities so that they are unaware that they might be engaged in illegal logging, fishing, and/or other activities. Participatory mapping is another approach used to engage community members, particularly fishers, in understanding the importance and effectiveness of no-take zones. Almost all of the campaigns use community patrols and monitor teams as a way to enforce the no-take zone, composed usually of volunteer teams (Wirawan, Kushardanto, & Tyas Nuhidayati, 2012).

Working within the paradigm of constructed potentiality, such groups offer two benefits to the campaign's goal of behavior change. First, these volunteers buy into the idea of the no-take zone. And second, they create a system of enforcement without requiring additional financial resources, demonstrating a move away from constricted potentiality. At the practical knowledge level, campaign managers have developed constructed potentiality by figuring out ways to work within the means that they have, often relying on community support and volunteers.

Another key strategy that campaign managers use is to develop a microcredit union within the local community of the campaign. Because the campaigns are almost always in rural and often isolated locations where fishers and their families subsist on the fish caught, finding ways to develop an income basis can be very valuable to community members. For example, in Komodo National Park in Indonesia, the campaign manager (Devi Opat) developed a credit cooperative in which community members agreed to participate in the no-take zone approach to conservation as well as to develop community savings (Rare Indonesia, 2015; Rare Indonesia, 2014). By putting small amounts of money into a community-managed credit cooperative, fishers were able to remove the need for middlemen or fish bosses, who often supplied fishing equipment and other capital to the local fishers but also took a big cut of their profits. When community members were able to save money and then purchase their own equipment, they were able to earn substantially more money without being in debt to these fish bosses, which then in turn decreased the need to fish (Rare Indonesia, 2014). Strategies like microcredit unions are successful in Indonesia because they meet and respond to the endemic economic and environmental factors that are contributing to problems like illegal fishing. These methods are driven by practical knowledge (and bound by the particular factors at work in specific geographic or cultural regions), but aim to empower publics by means of collective accountability. Although the microcredit union relies on material resources for its success, the creative implementation of this strategy suggests a constructive potentiality in which the community members became empowered and freed from the constricted potentialities of the fish bosses.

Local Knowledge

In Pride campaigns, local knowledge is an important symbolic resource for increasing public participation through constructed potentiality, as campaign managers create adaptive and responsive campaign strategies grounded in local understandings of culture, geography, politics, and social issues. CMs increase public participation by creating campaign materials and selecting mascots based on community input and choosing locations and activities for campaign events that are culturally significant for community members.

Rare's Indonesia office has also developed some campaign strategies around local knowledge. The program and Pride campaign development require extensive surveying of local community members to find out which social marketing strategies will work best in that particular community. While T-shirts with slogans and mascots and various school visits are successful in all campaigns, such materials are often locally produced. Slogans are often in local languages (there are more than 400 languages in Indonesia in addition to the primary Indonesian language that is taught in all public schools). As religion is a major part of life for almost all Indonesians, many campaign managers use local religious leaders, festivals, and ceremonies as a way to connect to communities. The role of traditional leaders and customs is also important in establishing community connection, participation and engagement. For example, in many Muslim areas, the campaign manager might involve the mosque and Friday prayers as part of the campaign strategy (Rare Indonesia, 2014). Boat races, using traditional long boats, are also popular in some campaigns, as are karaoke contests (Rare Indonesia, 2014; Wirawan, Kushardanto, & Tyas Nurhidayati, 2012). Many Indonesians also frequent food stalls known as warungs, so campaign managers have used stall and table banners as a way to connect with communities (Rare Indonesia, 2014). As cell phones are one of the key ways people communicate with each other in rural communities (and indeed, throughout the world), campaign managers have also implemented text message blasts, in which they send out conservation messages to community members who have signed up to receive such texts (Rare Indonesia, 2015). While many of the campaign managers use similar approaches, they all make extensive efforts to make sure that the campaign includes, engages, and emphasizes local community preferences, creating options for multiple, unspecified routes to change through constructed potentiality. As mentioned in the previous section, the system of volunteers has been instrumental for almost all the campaigns in Indonesia (and Malaysia and Timor Leste).

One significant way that Pride campaigns draw on local knowledge is through the use of campaign mascots based on local species specific to the geographic region. These mascots not only become the face of the campaign on promotional materials like T-shirts and stickers, but because they are designed based on feedback from local community members, they often become significant community figures. For example, in Mónica Andrea Rivera Suárez's campaign in La Unión, Colombia, local residents chose a sloth as their mascot. Because sloths used to live in a central park in the community, community members found it easy to generate sympathy for this species, and this local knowledge ultimately translated into support for the campaign. The emphasis on local knowledge means that Pride campaigns are able to draw from the experiences of individuals and groups who often are not included in conservation efforts. This concern for maintaining the voice and perspective of local stakeholders enhances outcomes because it frames participation "in terms they find resonant, important to them, thereby opening a portal into their communal standards for such action" (Carbaugh, 2005, p. xiii, quoted in Lindenfeld et al., 2012). Thus, the knowledge they bring to the table becomes an important symbolic resource for achieving meaningful environmental change. For example, Marco Bustamante's campaign to protect the Yanuncay watershed in Cuenca, Ecuador, faced difficulties because of a disconnect between the local community and the water company. A focus on local knowledge revealed that all the children in the community attended a school in Soldados, and this turned out to be a significant symbolic resource, as the school became a central meeting place for Bustamante's campaign. His campaign held community events like puppet shows for the children at the school, and this site became an important place for parents to hear campaign messages.

Perhaps one of the most significant examples of local knowledge in Latin American Pride campaigns is the use of music at community events to raise awareness. Every Pride campaign composes a song as part of its social marketing strategy, but these songs have a special significance in Latin American campaigns. Campaign managers use popular dance music, like salsa, cumbia, and bachata, to enable campaign messages to be shared at community events where dancing plays a large role. These examples of local knowledge demonstrate how thinking strategically and creatively about symbolic resources and constructed potentialities can have a big impact within a community, particularly when extensive efforts are made to use appropriate cultural values and symbols.

Conclusion

Pride campaigns in Indonesia, Latin America, and China have successfully used institutional, practical, and local knowledge as a symbolic resource to achieve significant conservation results. This focus has allowed Pride campaigns to create change through the paradigm of constructed potentiality. Constructed potentiality creates multiple, unspecified routes to change, and Pride campaigns have successfully used various strategies as routes to change. For example, in several of the Latin American campaigns discussed above, campaign managers worked with community members to create spaces where upstream land owners and downstream water users could make time to meet and communicate with one another. This was ultimately a simple yet effective change strategy because it was the first time many residents were able to express their concerns and relate to one another over their shared use of water. In China, cultural norms proved to be an important part of the route to change, as family-oriented values guided decision-making processes for fishermen. Indonesian campaigns, for example, have used volunteer programs to create both buy-in from local communities as well as reduce the need for economic resources.

Constructed potentiality achieves change by channeling efforts internally, and this leads to opportunities for self-change. Through public participation in Pride campaigns, the internal focus of change efforts has allowed stakeholders at multiple levels to engage in self-change, consistently redefining their relationship to the environment and their roles in their communities and using self-reflection as a tool for growth. In Indonesia, key stakeholder meetings that created participatory mapping and volunteer community patrols are examples of internal change. Campaigns in Latin America have created spaces for community members to come together and create mutually beneficial agreements to protect watersheds, illustrating public participation and opportunities for self-change. In China, the slogan "Protecting the Environment Will Protect Our Descendants" allowed fishermen to see themselves as part of the process of environmental conservation, and choosing to change their behaviors became a way to practice family values.

Not only do these campaigns address self-change through constructed potentiality, but they also shape frameworks for government policies and regulatory practices at local and national levels. For example, many of the campaigns in Indonesia have worked with local governments to establish marine protected areas or zones; in some cases, Rare's advocacy has expanded to the national government of Indonesia. That is, the success of Rare's campaigns around marine protected areas has led to national-level discussions and regulatory frameworks, at least in part, on the part of the national park system (under the authority of the Ministry of Forestry). In addition, these campaigns are designed for practical implementation through SMART (Specific, Measurable, Achievable, Realistic, Timebound) objectives. They are then assessed rigorously through knowledge-attitude-practice (KAP) surveys and other methods, such as biodiversity measures to assess fish stock populations in marine protected areas, for example.

In this chapter, we have argued that Rare's use of knowledge as a symbolic resource enables a form of constructed potentiality that can bolster public participation outcomes and enhance governmental regulatory and decision-making processes. Through Rare's reliance on institutional, practical, and local knowledge, these campaigns embrace the paradigm of constructed potentiality. This approach to public participation means that rather than experts relying on persuasion as a tool to demand change from community members, Rare's efforts allow individuals to see themselves as part of the change process. Through their own choice to change their relationships to the surrounding environment, local residents are empowered to take part in conservation. The resulting pride in local conservation means that campaigns can achieve lasting, sustainable solutions to conservation challenges.

References

Alger, K. (2014). Reciprocal water agreements for watershed protection. Washington, DC: Rare. Retrieved from https://blog.nationalgeographic.org/2014/06/17/reciprocal-water-agreements-for-watershed-protection/

Barge. J. K., & Schockley-Zalabak, P. (2008). Engaged scholarship and the creation of useful organizational knowledge. *Journal of Applied Communication Research, 36*(3), 251–265.

Barley, W. C., Leonardi, P. M., & Bailey, D. E. (2012). Engineering objects for collaboration: Strategies of ambiguity and clarity at knowledge boundaries. *Human Communication Research, 38,* 280–308.

Burns, T. R., & Uberhost, R. (1988). *Creative democracy: Systematic conflict resolutions and policymaking in a world of high science and technology.* New York: Praeger Press.

Callister, D. C. (2013). Land community participation: A new "public" participation model. *Environmental Communication, 7*(4), 435–455. doi:10.1080/175240 32.2013.822408

Canary, H. E. (2010). Structurating activity theory: An integrative approach to policy knowledge. *Communication Theory, 20,* 21–49.

Canary, H. E., & McPhee, R. D. (2011). Introduction: Toward a communicative perspective on organizational knowledge. In H. E. Canary & R. D. McPhee (Eds.), *Communication and organizational knowledge: Contemporary issues for theory and practice* (pp. 1–14). New York, NY: Routledge.

Carbaugh, D. A. (2005). *Cultures in conversation.* Mahwah, NJ: Lawrence Erlbaum.

Cox, R. (2004). "Free trade" and the eclipse of civil society: Barriers to transparency and public participation in NAFTA and the free trade area of the Americas. In S. Depoe, J. Delicath, & M. A. Elsenbeer (Eds.), *Communication and public participation in environmental decision making* (pp. 201–219). Albany, NY: State University of New York Press.

Daniels, S. E., & Walker, G. B. (2001). *Working through environmental conflict: The collaborative learning approach.* Westport, CT: Praeger.

Dulipovici, A., & Baskerville, R. (2007). Conflicts between privacy and property: The discourse in personal and organizational knowledge. *Journal of Strategic Information Systems, 16*(2), 187–213.

Foss, S. K., & Foss, K. A. (2011). Constricted and constructed potentiality: An inquiry into paradigms of change. *Western Journal of Communication, 75*(2), 205–238.

Foss, S. K., & Foss, K. A. (2012). *Inviting transformation: Presentational speaking for a changing world* (3rd ed.). Long Grove, IL: Waveland Press.

Foss, S. K., & Griffin, C. L. (1995). Beyond persuasion: A proposal for an invitational rhetoric. *Communication Monographs, 62*(1), 2–18.

Gorsevski, E. W. (2004). *Peaceful persuasion: The geopolitics of nonviolent rhetoric.* Albany, NY: State University of New York Press.

Hodgkinson, G. P., & Healey, M. P. (2008). Cognition in organizations. *Annual Review of Psychology, 59,* 387–417.

Kuhn, T., & Porter, A. J. (2011). Heterogeneity in knowledge and knowing: A social practice perspective. In H. E. Canary & R. D. McPhee (Eds.), *Communication and organizational knowledge: Contemporary issues for theory and practice* (pp. 17–34). New York, NY: Routledge.

Lakoff, G. (2010). Why it matters how we frame the environment. *Environmental Communication: A Journal of Nature and Culture, 4*(1), 70–81.

Lam, A. (2000). Tacit knowledge, organizational learning and societal institutions: An integrated framework. *Organization Studies, 24,* 116–138.

Leonardi, P. M. (2011). Information, technology, and knowledge sharing in global organizations: Cultural differences in perceptions of where knowledge lies. In H. E. Canary & R. D. McPhee (Eds.), *Communication and organizational knowledge: Contemporary issues for theory and practice* (pp. 89–112). New York, NY: Routledge.

Lindenfeld, L. A., Hall, D. M., McGreavy, B., Silka, L., & Hard, D. (2012). Creating a place for environmental communication research in sustainability science. *Environmental Communication, 6*(1), 23–43.

McPhee, R. D., Corman, S. R., & Dooley, K. (1999). *Theoretical and methodological axioms for the study of organizational knowledge and communication.* Unpublished manuscript.

Murphy, A. G., & Eisenberg, E. M. (2011). Coaching to the craft: Understanding knowledge in health care organizations. In H. E. Canary & R. D. McPhee (Eds.), *Communication and organizational knowledge: Contemporary issues for theory and practice* (pp. 264–284). New York, NY: Routledge.

Perrow, C. (1986). *Complex organizations: A critical essay* (3rd ed.). New York, NY: McGraw-Hill.

Rare (n.d.–a). Pride campaign. Washington, DC. Retrieved from http://www.rare.org/pride#.VqbhdNIrLIV.

Rare (n.d.–b). *The Rare approach: Community-based solutions for global conservation.* Washington, DC: Rare. Retrieved from https://www.rare.org/sites/default/files/RareApproachWEB.pdf

Rare (2014). Using social marketing to protect biodiversity. 73 campaigns in 2010 alone. Washington, DC: Rare. Retrieved from https://www.rare.org/en-press-global-journal-article-2012#.WV9OxojyvIU

Rare (2015). Get to know the team. Washington, DC: Rare. Retrieved from http://www.rare.org/campaign/china-sustainable-ag

Rare (2016a). *Good ideas abound: Motivating people to adopt them is rare.* Washington, DC: Rare. Retrieved January 30, 2016 from https://www.rare.org/about

Rare (2016b). *Where Rare works: Conservation solutions in China.* Washington, DC: Rare. Retrieved from http://www.rare.org/places

Rare Indonesia (2014). *Dua belas langkah awal untuk lompatan besar.* Bogor, Indonesia: Rare.

Rare Indonesia (2015). *Rare Indonesia dalam konservasi kelautan dan perikanan berkelanjutan.* Bogor, Indonesia: Rare.

Ryan, K. J., & Natalle, E. J. (2001). Fusing horizons: Standpoint hermeneutics and invitational rhetoric. *RSQ: Rhetoric Society Quarterly, 31*(2), 69–90.

Senecah, S. L. (2004). The trinity of voice: The role of practical theory in planning and evaluating the effectiveness of environmental participatory processes. In S. P. Depoe, J. W. Delicath, & M. A. Elsenbeer (Eds.), *Communication and public participation in environmental decision making* (pp. 13–33). Albany, NY: State University of New York Press.

Waddell, C. (1996). Saving the Great Lakes: Public participation in environmental policy. In C. G. Herndl & S. C. Brown (Eds.), *Green culture: Environmental rhetoric in contemporary America* (pp. 141–165). Madison, WI: University of Wisconsin Press.

Wirawan, N. P. S., Kushardanto, H., & Tyas Nurhidayati, A. (2012). *Mengasah batu menjadi erlian: Belajar dan bekerja bersama komunitas untuk menciptakan perubahan.* Bogor, Indonesia: Rare.

Yang, T. (2015). *Environmental problems in China.* Washington, DC: World Wide Fund For Nature (WWF).

Chapter 10

Fracking, the Elsipogtog First Nation, and Disruptive Public Participation

The Role of Images in Amplifying Outrage on Twitter

MOLLY SIMIS-WILKINSON AND JILL E. HOPKE

Introduction

Much has changed in the landscape of civic participation—across sectors from political campaigning to environmental and social justice movements—since the advent of social media. The Arab Spring revolutions, the Spanish *Indignados* (or the "Outraged"), and Occupy's takeovers of public squares around the world in 2011 to the Black Lives Matter movement's calls for accountability in policing and resistance to the Dakota Access Pipeline in 2016 show the changing dynamics of public participation. In Canada, the Indigenous peoples' Idle No More uprising captured the attention of the public and spread internationally from late 2012 onward. The dynamics of public participation are changing toward more disruptive forms of taking and controlling spaces, both physical and digital. The 2013 Elsipogtog First Nation protests and subsequent police actions present an opportunity to study the dynamics of environmental conflict and the role of visual social media posting as a form of "disruptive public participation." In this case, that refers to the practice of circulating images on Twitter with the intent to express and amplify a sense of public outrage. We consider *disruptive public participation*[1] to involve constituents who perceive themselves as outsiders to the decision-making processes or who are from historically marginalized groups and who make use of digital and social media applications to amplify and document dissent when traditional modes of public participation have been exhausted or are perceived to be ineffective to the achieve the goals of stakeholders.

In the fall of 2013, members of the Elsipogtog First Nation staged a blockage against shale gas exploration in New Brunswick, Canada. Eventually, the Royal Canadian Mounted Police (RCMP) used force to remove peaceful protesters. On October 17, 2013, the RCMP raided the Mi'kmaq Warrior camp holding an anti-fracking blockade along Route 134 near Rexton, New Brunswick (CBC News Staff, 2013). The blockade had been going on since September 30, 2013, when protesters moved into the area to block a fracking company's equipment staging area just off Route 134 (CBC News Staff, 2013). On that date, RCMP moved in, with officers pepper spraying and arresting dozens of protesters. Throughout the day, Molotov cocktails were thrown at the police, five police cars were set on fire—ostensibly by protesters—and RCMP snipers staked out the site and drew their weapons (CBC News Staff, 2013; Howe, 2013).

Dramatic images of these events circulated in real time on the social media platform Twitter as they unfolded and in the days afterward. The anti-shale demonstrations took place within the context of the Idle No More movement for Indigenous rights. Idle No More is an activist movement of Indigenous, First Nations peoples throughout Canada advocating for Indigenous rights and equality. Many Twitter posts about the Elsipogtog First Nation blockade were cross-hashtagged with the Idle No More movement hashtags #idlenomore and #INM.

Disruptive public participation by marginalized stakeholders vying for their voices to be heard using social media raises new questions for environmental public policy, most notably how these applications are providing new digital spaces for civic engagement on, and public participation in, decision making over natural resource extraction. Using the Elsipogtog First Nation anti-shale development protests and subsequent police actions as a case study, we examine public participation in an environmental conflict through analyzing *amplification effects* on Twitter. By *amplification effects*, we mean how wide reaching social posts are as indicated by the number of retweets. Furthermore, we also consider what happens in cases in which dissenting voices are likely be suppressed in public participatory practices, as more conventionally defined. Traditionally, environmental movements focus their efforts on alerting, amplifying, and rallying broader publics to their causes (see Cox & Pezzullo, 2016).

Through an illustrative case study and associated empirical analysis, we consider how the social media platform Twitter might facilitate approaches to disruptive public participation in decision making over natural resource extraction rooted in participatory democracy. In practice, it is not clear that social media applications are constitutive spaces for "productive dialogue" between stakeholders with varying backgrounds of "technical expertise" and "local knowledge" (Depoe & Delicath, 2004, p. 10; Hopke & Simis, 2017). Exploring the discourse and imagery on social media around this contentious topic is important to understanding how dissonance manifests on social media. Understanding this dissonance means understanding the lens through which publics and environmental decision makers may interpret the course of events as they use social media to inform or promote their own perspectives.

Public Participation in an Era of Social Media

In the introduction to *Communication and Public Participation in Environmental Decision Making*, Depoe and Delicath (2004) ask two fundamental questions about community participation in environmental decision making dealing with the role for individuals in generating and applying environmental policy and ways to "maximize value and impact of public input" (2004, p. 1). They go on to distinguish between public participation that is "within" institutional frameworks and modes of participation that fall "outside" these frameworks (2004, p. 1). An iconic and oft-cited representation of public participation in the public sphere is Arnstein's (1969) ladder of citizen participation.[2] The ladder has eight rungs, with the bottom rungs of (1) manipulation and (2) therapy constituting nonparticipation; the middle rungs of (4) information, (5) consultation, and (5) placation constituting tokenism; and the highest rungs of (6) partnership, (7) delegated power, and (8) citizen control constituting citizen power (Arnstein, 1969). Though useful, the separation between the rungs is fluid and could be broken down into smaller increments in real-world contexts (Arnstein, 1969). Additionally, this ladder of citizen participation oversimplifies the relationship between citizens and other power holders. Indeed, one of the limitations of the ladder is that socioeconomic and demographic obstacles to participation are not adequately addressed (Arnstein, 1969). An explicitly more fluid ladder that attempts to address some of these shortcomings includes the bottom rungs of (1) education, (2) information feedback, and (3) consultation, all actions taken by the general public (Connor, 1988). The top rungs are then actions taken by leaders, including (4) joint planning, (5) mediation, (6) litigation, and (7) resolution/prevention (Connor, 1988). Citizen participation in this framework is cumulative, with the rungs successively building on each other (Connor, 1988).

Public participation in both of these ladders is in some ways restrictive. Citizens with critical and alternative views face challenges to engaging in decision-making processes, as discussed by Hunt et al. in their contribution to this volume. There is little to no trustworthy mechanism for engagement of dissenting voices, particularly as a commonly used mechanism—soliciting public comments—has been shown to be "discursively managed" by decision makers in a way that highlights supportive citizen voices to justify the proposed action and that buries critical voices (Carvalho, Pinto-Coelho, & Seixas, 2016, p. 15). Newer modes of citizen participation in environmental policy and decision making engaged through "digital mediated social networks"—applications such as Facebook, Twitter, and Tumblr—are not easily plottable on the ladder discussed above (see Cox & Pezzullo, 2016, p. 215). Citizens and activists use social media to engage with each other, with public audiences, and with stakeholders toward a range of goals, from increasing awareness of an issue and amplifying its presence on a global stage to calling for an organizing action (Cox & Pezzullo, 2016). Digital media create a new public sphere that serves as a "hybrid space" blurring the boundaries of online and offline political action (Callison & Hermida, 2015). Digital media provide outlets for critics of decision-making processes and other dissenting voices to be heard when they might otherwise be excluded from meaningful formal engagement. For environmental social movements, digital and social media networks can serve three functions: to "alert" publics to an issue, to "amplify" "visual testimony" of events, and to provide spaces to "engage" in protest (Cox & Pezzullo, 2016, pp. 210–213).

Nabatchi and Leighninger (2015) offer a broader, more inclusive definition of public participation as "an umbrella term that describes the activities by which people's concerns, needs, interests, and values are incorporated into decisions and actions on public matters and issues" (p. 14). We argue that Elsipogtog activists and their supporters used Twitter to amplify a movement for social change that is implicit in this definition of public participation. Through our case study, we explicate how this seemingly unconventional type of public participation expands the boundaries of the traditional framework to include *disruptive* public participation.

Through examining the discourse and visual imagery on Twitter about the Elsipogtog protests and police action crackdown, we study the social media presence of this environmental justice movement, in-the-moment reactions to the RCMP raid, and the potential amplification effect of photographs and other visual imagery in social media. Recent research related to Indigenous social justice movements on social media have considered the actors involved in the discussion (Callison & Hermida, 2015) but not the content of messages. Social media platforms, including Twitter, allow users to remediate images and visuals (e.g., photographs) for their own purposes, which can aid in their going "viral" and the "multiplication" of messages (Cox & Pezzullo, 2016, p. 214). Additionally, research related to the amplification of images on Twitter is limited to examining a subset of influential and favorited posts (Chung & Yoon, 2013) but does not include an assessment of the role of visual images across a corpus of tweets for a time-bound event that has a strong presence on the medium. In this chapter, we argue that this unconventional, uninvited form of disruptive public participation expands the boundaries of the traditional public participation theoretical framework.

From #IdleNoMore to #Elsipogtog: Bringing Together Indigenous Voices on Twitter

Starting in 1701, the British government made a series of peace and friendship treaties with the First Nations tribes of the Maritimes region (Mulrennan, 2015). Much has changed in recent decades in the way that both the federal and provincial Canadian governments recognize the "collective rights" of Indigenous peoples (Mitchell, 2015, p. 16). A 1996 Report on the Royal Commission on Aboriginal Peoples outlines four stages in relations with the Indigenous tribes of the territory that is now Canada: 1) "separate worlds (pre-contact)," 2) "contact and co-operation (contact to 1812)," 3) "displacement and assimilation (1812 to early 1970s)," and 4) "negotiation and renewal (early 1970s to the present)" (Mulrennan, 2015, pp. 59–62). The 1982 Canada Act specified that provincial governments control natural resources on crown lands that are not privately owned with "legislative authority" shared between the provincial and federal levels of government (Mitchell, 2015, pp. 15–16). Under the 1982 Canada Act, "existing Aboriginal and treaty rights were formally recognized" (Mitchell,

2015, p. 16). The Canadian government, at both federal and provincial levels—and recognized under international law with the 2007 United Nations Declaration on the Rights of Indigenous Peoples (UNDRIP)—has a "duty to consult and accommodate" Indigenous communities on issues that could "infringe" on their treaty rights (Mulrennan, 2015, pp. 55–56).

In late 2012, four women in the Canadian province of Saskatchewan organized a teach-in about a then-proposed federal omnibus budget bill C-45. The spending bill removed environmental protections for waterways and diminished First Nations autonomy over land management (Guertin & Buettner, 2014; Sinclair, 2014). The women called the action "Idle No More" as a call to action to "to get off the couch and start working," according to one of the founders (Caven, 2013). Through the end of 2012 and into 2013, the grassroots, horizontally networked movement—meaning contributors connected with each other and bypassing traditional social movement organizational ties or formal hierarchy, what Hopke (2016) calls a "translocal" movement structure—grew into the most widespread Canadian social justice movement seen after the 1960s (Sinclair, 2014). The movement gained widespread media attention, including of a more than a monthlong hunger strike by Attawapiskat Chief Theresa Spence, who camped in a tipi adjacent to the Canadian Parliament in an effort to gain an audience with then Prime Minister Stephen Harper (Callison & Hermida, 2015). While the movement started in response to Harper's conservative government's "failure to consult Aboriginal peoples" on bill C-45, the Jobs and Growth Act, a piece of legislation with repercussions for Indigenous communities and treaty rights, this women-led Indigenous movement has grown into a more far-reaching call for "collective cultural resurgence" (John, 2015; McMillan, Young, & Peters, 2013, p. 430).

The traditionally defined means of public participation—for example, the rungs of Arnstein's (1969) ladder, from information and consultation to citizen power—are not always readily available to Indigenous communities. In the case of Elsipogtog, the community was directly affected, but the mechanisms to participate in decision-making processes were not clear. This means that these Indigenous communities had to insert themselves as stakeholders by finding means to impact decision making outside conventional political channels to build grassroots power and influence—what we term *disruptive public participation*.

As the movement grew, the phrase "Idle No More"—and the Twitter hashtag of #IdleNoMore—developed into a powerful counternarrative that changed the course of the mainstream political dialogue in Canada (Guertin & Buettner, 2014). Furthermore, the Idle No More movement grew from the grassroots in Indigenous communities rather than through First Nations leadership structures (Callison & Hermida, 2015). Much scholarship in political sociology supports the idea that networks of social ties can act as "conduits for activism" (Gould, 2003, p. 254). In the case of Idle No More, Wood (2015) finds the movement diffused, in part, through activists' use of social networking and media platforms, with shared themes of solidarity, resurgence, and Indigenous sovereignty.

While the Canadian spending bill C-45 ultimately passed into law, Idle No More brought together Indigenous peoples from around the world into a common banner and shared collective identity. Activists held several days of action, the first on International Human Rights Day, December 10, 2012 (Guertin & Buettner, 2014). The height of the movement, as measured by #IdleNoMore Twitter post volume, was December 2012 through January 2013 (Callison & Hermida, 2015). In a study of #IdleNoMore Twitter messages during this time period, Callison and Hermida (2015) find that Twitter functions as a "hybrid space," providing a platform for contesting narratives, the development of a shared collective identity, and crowdsourcing Indigenous voices into elite discourse, which is dominated by institutional actors such as journalists, politicians, and celebrities. Non-elites are represented among the top 500 most retweeted and influential Twitter users talking about Idle No More, but the top 25 most influential accounts are made up of by and large "institutional elites," including journalists and celebrities (Callison & Hermida, 2015, p. 697). They find, however, that of the top most retweeted, or amplified, voices, there is a higher representation of Indigenous individuals (Callison & Hermida, 2015). Indigenous activists, like social movement organizers and supporters across varied causes, have taken up creative applications of digital media tools to amplify transnational interlinkages and build solidarity (Dreher, McCallum, & Waller, 2016).

The crowdsourced and networked #IdleNoMore frames in the early months of the movement centered on colonialism, building alliances, and calling for social change (Callison & Hermida, 2015). Idle No More activists used Twitter to make "strategic alliances" with other movements (Callison & Hermida, 2015), such as Elsipogtog First Nation. These results are supported by research into the transnational anti-fracking movement Global Frackdown's Twitter practices during a 2013 global day of action against fracking (Hopke, 2015). Hopke (2015) finds that Global Frackdown tweeters framed anti-shale activism largely in terms of solidarity and convergence between social movements with shared, or overlapping, objectives and collective identities, including Idle No More.

First Nations communities in Eastern Canada, including the Mi'kmaq, have been organizing for decades to protect their treaty rights and have spent centuries in resistance to colonialism (see McMillan et al., 2013). The traditional territory of Mi'kmaq First Nation consists of seven districts along Canada's border with the Atlantic Ocean and the maritime provinces east of the St. John River, which includes Prince Edward Island, Nova Scotia, eastern New Brunswick, and Quebec's Gaspe region ("Mi'kmaq," 2015a; "Mi'kmaq," 2015b). Within the context of the broader Idle No More movement, in June 2013 the Elsipogtog First Nation, a band in the Sigenigteoag Division of Mi'kmaq First Nation in New Brunswick, put out a call for support of its Sovereignty Summer Campaign to "prevent seismic testing vehicles and workers from testing for shale gas deposits for purposes of resource exploitation on Indigenous territories" (Elsipogtog First Nation, 2013, June 24; "Mi'kmaq," 2015b).

The New Brunswick Oil and Natural Gas Act allows for the "seamless transition" from an exploration license to an extraction production one (Howe, 2015, p. 50). Additional public consultation is not required (Howe, 2015). The government of New Brunswick had held a "Provincial Forum on Natural Gas Development" in June 2011 (Howe, 2015). The premier of New Brunswick at the time, David Alward, of the Progressive Conservative Party of New Brunswick, decided that seismic testing by Southwestern did not fall under the Duty to Consult with Aboriginal peoples (Howe, 2015, p. 75). However, Southwestern Energy Company, in conjunction the Assembly of First Nations Chiefs of New Brunswick Inc. (AFNCNB), did host a series of community meetings and presentations with First Nations, including Elsipogtog, from May 2012 through September 2013 (Howe, 2015). According to Howe (2015), the "AFNCNB consultation process work had more to do with community infiltration, propaganda dispersal and company brand management" (p. 75).

The First Nation demonstrators faced a police crackdown that resulted in the arrest of several dozen activists, including a dozen on Aboriginal Day (Elsipogtog First Nation, 2013, June 24). Activism against shale development in New Brunswick continued into the fall. On October 17, 2013, the dispute over shale gas exploration in the region came to a head. The Royal Canadian Mounted Police (RCMP) moved in during a predawn raid to enforce an injunction by Southwestern Energy, a company conducting seismic testing in the area, against a blockade by members of the Elsipogtog Mi'kmaq First Nations tribe (Howe, 2013). Members of the Elsipogtog First Nation had been contesting the shale project through litigation and direct action since that summer. The RCMP arrested more than 40 people on October 17, 2013, when enforcing the injunction against the Elsipogtog blockade. Contention over shale exploration in New Brunswick had ongoing for several months prior to an international day of action against fracking, called the Global Frackdown, which took place on October 19, 2013 (see Hopke, 2015; Hopke & Simis, 2017).

As noted by journalist Miles Howe, who covered the October 17, 2013 RCMP raid firsthand and was himself arrested that day, "None of Canada's Maritime provinces are ceded land. The Crown is tied to the original Indigenous inhabitants—and their land—through treaties of peace and friendship" (Howe, 2013, para. 20). In an op-ed in support of the Elsipogtog appearing in November 2013 in the *Toronto Star*, Métis writer and educator Chelsea Vowel summarized the situation in New Brunswick:

> You have a group of people who never gave up ownership of their land or resources, opposing widely contested shale gas exploration, which was approved by a government that does not own the land or resources, acting with the support of their non-native neighbours and being reported on by mainstream media outlets that often fail to address the substantive issues. (Vowel, 2013, para. 10)

Images of the October 17, 2013, RCMP raid on the Elsipogtog blockage quickly spread around the world, bringing support in terms of material resources, supporters from as far away as Western Canada, and international solidarity (Howe, 2015). Activists and supporters used both established #IdleNoMore and #INM hashtags as well as #Elsipogtog, #cdnpoli (short for "Canadian politics"), and #rcmp to share information and images of the blockage and arrests. Activists can use "well established" hashtags, such as #IdleNoMore, to shortcut around mainstream media and broadcast their messages directly to supporters and the public (Callison & Hermida, 2015, p. 696). In the face of public outrage across Canada and internationally by the end of 2013, Southwestern Energy Company withdrew from New Brunswick, stating in a press release on December 6, 2013, "SWN Resources Canada is pleased to announce that we have successfully completed our seismic acquisitions program" (as cited in Howe, 2015, p. 171). The public pressure on the industry stakeholders was facilitated by social media activity, which empowered and amplified marginalized voices. We offer empirical support for this claim through an analysis of visual and textual Twitter messages across levels of amplification. We argue that the disruptive ways stakeholders engage in decision-making processes expand the boundaries of the traditional public participation framework.

Analyses

Having overviewed the relevant literature and case background, we turn our attention to the results of our analyses of the discourse and images on Twitter about the Elsipogtog protests and associated police actions for a two-week time period in October 2013. We explore the portrayal of these interactions on Twitter following the RCMP October 17, 2013, raid on the Elsipogtog encampment. We use a data set of several thousand tweets collected during the protests to investigate features of those that are particularly wide-reaching compared with dead-end tweets that get little to no attention. Additionally, we compare posts that have images with those that do not. Disruptive public participation operates outside the margins of mainstream discourse, perhaps promoting solidarity among traditionally excluded groups. We speculate that the social media presence of the Elsipogtog resistance galvanized people with images that provoked moral outrage, turning public attention to an issue that had been going on for some time.

We qualitatively analyze the images associated with the most highly retweeted posts to further describe the police-protester encounters depicted. We apply a "descriptive coding" method, which Saldaña suggests for a range of data types, from text to imagery (2009, pp. 70–73). He writes that this coding method "leads primarily to a categorized inventory, tabular account, summary, or index of the data's contents" (Saldaña, 2009, p. 72). In analyzing the data qualitatively, we attempt to ground our research in "a larger historical, cultural, and global context" (Chilisa, 2012, p. xv). Through processes of self-reflection, we seek to acknowledge Indigenous knowledge systems outside the Western social scientific tradition in which "knowledge is relational" rather than an individualistic pursuit (Chilisa, 2012, p. 21).

TWITTER TEXTUAL ANALYSIS

The majority (63%) of U.S. adults on Twitter turn to the platform as a source of news (Pew Research Center, 2015a). When tweeting about news, users were more likely to retweet than to post original content, a pattern that Pew Research Center (2015b) observed is the opposite of posting about topics other than news. Thus, Twitter is a useful social media outlet to focus on, as users are increasingly turning to the medium for news-related purposes, as one would in the face of a breaking news story like that of the Elsipogtog First Nation protests (Pew Research Center, 2015a, 2015b).

Data collection and data set creation. To collect and analyze Elsipogtog Twitter posts from the time of the RCMP raid, we began with a data set of English-language, fracking-related tweets, collected via cloud-based data analytic software DiscoverText. We generated the data set by gathering all posts that used one of the following five hashtags between October 13 and October 27, 2013, from the Twitter firehose: #fracking, #globalfrackdown, #natgas, #shale, and #shalegas (N = 64,973).[3] Hashtags, denoted by the "#" sign and called "searchable talk" by Zappavigna (2012), are a way of cataloging conversations on Twitter (p. 95). Twitter users use hashtags as parts of their posts and as a way to keep track of topics of interest to them. This time frame covers one week preceding and one week following the protests.

We narrowed down these fracking-related tweets with the following hashtag keywords: #cdnpoli, #rcmp, #Elsipogtog, #INM, #IdleNoMore, #nbpoli, #FirstNation, and #FN. Then we used DiscoverText's de-duplication function to remove verbatim repeat posts from the data set. The de-duplicating process compares strands of text (meaning words that are next to each other) and groups posts that have a 100% exact match. These exact matches are reasonably assumed to be retweets of an original post. We created our working data set out of the nonduplicate posts (n = 3, 567). We then sorted these groups into three categories based on the number of retweets for each post: high (20+ retweets), low (1–19 retweets), and single items (no retweets). We further sorted these three categories into *image* and *no image* groups, giving us six final data sets to analyze (see Table 10.1).

Table 10.1. Information about the makeup for each of the six data sets

Retweet category (total size)	Image/No image	Number of coded items
High # of retweets (20+) (n_h=40)	Image	20
	No image	20
Low # of retweets (1–19) (n_l=761)	Image	242
	No image	519
Single items (no retweets) (n_s=2,872)	Image	703
	No image	258

Content analysis. We grounded our coding schema in recent work on police and protestor interaction on social network sites (Earl, McKee Hurwitz, Mejia Mesinas, Tolan, & Arlotti, 2013) as well as inductively determined themes by iteratively examining the posts to ensure that the coding schema was exhaustive for our data set (see Table 10.2). We coded every post (image included, when applicable) on two dimensions: "movement response/ presence" (including peaceful protest, disruptive protest, and solidarity) and "police action/presence" (including use of force, surveillance and monitoring, and police equipment and weapons) (Krippendorff's alpha > 0.80). Across the six categories, we coded 1,767 posts. For all but the final category, the complete set was coded. For the final category of single items with no image, we coded a random sample consisting of 20% of the total posts in that category (see Table 10.2).

Table 10.2. Coding schema for textual analysis of Twitter posts related to Elsipogtog First Nation protests from October 13 to October 27, 2013. Handles (names of users, denoted by an "at" symbol, or "@") who are not public officials or public figures are redacted.

Dimension	Code name	Code definition	Example
Movement response / presence	Peaceful protest	Tweet mentions peaceful protests, including mention of blockade action	This is nuts: RT @[redacted] The legitimate response to peaceful protest in Canada? http://t.co/dxsIGSrwBn #Elsipogtog #cdnpoli #fracking
	Disruptive protest	Tweet mentions disruptive protests or protesters who are in physical confrontations with the police	RT @[redacted]: yeah the sub group of protesters were Agent Provocateurs.#cdnpoli #elsipogtog #Fracking #PnPCBC #StopHarper #I
	Solidarity	Tweet mentions protesters' solidarity or other individuals/groups expressing solidarity with protesters	With #OpFrackOff, #Anonymous lends support to Canadian protestors http://t.co/K0Q7Z14Zd6 #News #fracking #Mi'kmaq #water #land

Dimension	Code name	Code definition	Example
	No movement response / presence	Tweets do not mention any movement response	An opinion on @[redacted]'s nuanced pro #fracking alignment...#nbpoli #Elsipogtog @[redacted]
Police action / presence	Use of force / surveillance and monitoring	Tweet includes mentions of using force, arresting protesters, conducting raids, provoking protests, presence of guns, monitoring protesters, restricting the movement of protesters, dispensing protesters	RT @[redacted]: Harper dictatorship deploys armed goons to disperse indigenous protesters http://t.co/E4coqtyYZe #cdnpoli #Canada
	Equipment and weapons	Tweet mentions specific types of police equipment (e.g., variation in kinds of police vehicles, riot gear, barricades) or weapons (e.g., tear gas, batons, pepper spray)	RT @[redacted]: RCMP cars on fire at #fracking blockade in NB. #idlenomore http://t.co/BXZj4Y5WKl
	No police action / presence	There is no mention of police presence or influence	The Truth! @[redacted]: @[redacted]'s essay: Another Story from #Elsipogtog http://t.co/RJtAcm1lCC via @[redacted] #fracking #Elsipogtog

Findings. We conducted two chi-square tests for independence using the statistical software package SPSS version 20. We first compared the independent variable of "post type" with the dependent variable of ".movement response / presence." Second, we compared the independent variable of "post type" with the dependent variable of "police action / presence." The data were weighted by count. We examined Cramer's V test statistics to test the strength of the relationships. A higher value signifies a stronger association between the two variables (Hayes, 2005).

For the "movement response/presence" coding dimension, there are significant differences across all six categories (Pearson's Chi-square = 244.96, p-value <.0001; Cramer's V = .216). "Peaceful protest" is a dominant theme of the highly retweeted posts. Additionally, "solidarity" tends to be textually described rather than visually depicted with images and photographs. We found that "police use of force/surveillance and monitoring" was more often visually depicted with photographs and images than textually described.

For the "police action/presence" coding dimension, there are significant differences across all six categories (Pearson's Chi-square = 206.97, p-value <.0001; Cramer's V = .243). In other words, posts that have been highly retweeted with an image are significantly different from posts that have been highly retweeted without an image, and so forth. The pattern of "police action/presence" codes were similar among the single items and low # of retweets groups, while the codes varied in the high # of retweets volume group.

One substantial difference is that "no police action" constitutes the highest proportion of the three codes in the "no image" category for the low # of retweet and single items volume groups, while "use of surveillance and monitoring" constitutes the highest proportion in the "no image" category for the high number of retweets. Text-based posts with "no police action" mentioned are less frequently retweeted than text-based posts that mention "police use of force."

IMAGE ANALYSIS

Quantitative findings. We coded every image for the presence or absence of "police-protester interaction." The presence of "police-protester interaction" is associated with a higher numbers of retweets (see Figure 10.1). We conducted a chi-square test for independence using the statistical software package SPSS version 20. We compared the independent variable of "number of retweets" with the dependent variable of "police-protester interaction." As with the previous tests, the data were weighted by count. We additionally examined Cramer's V test statistics to test the strength of the relationships.

Though there appear to be significant differences among high, low, and single-item categories that have "police-protestor interaction" (Pearson's Chi-square = 7.791, p-value <.05), there is very weak association between the variables (Cramer's V = .086). This is likely due to the low sample size. Despite the weak statistical association, these data preliminarily point to images of interactions between police and protesters constituting a higher proportion of highly retweeted posts.

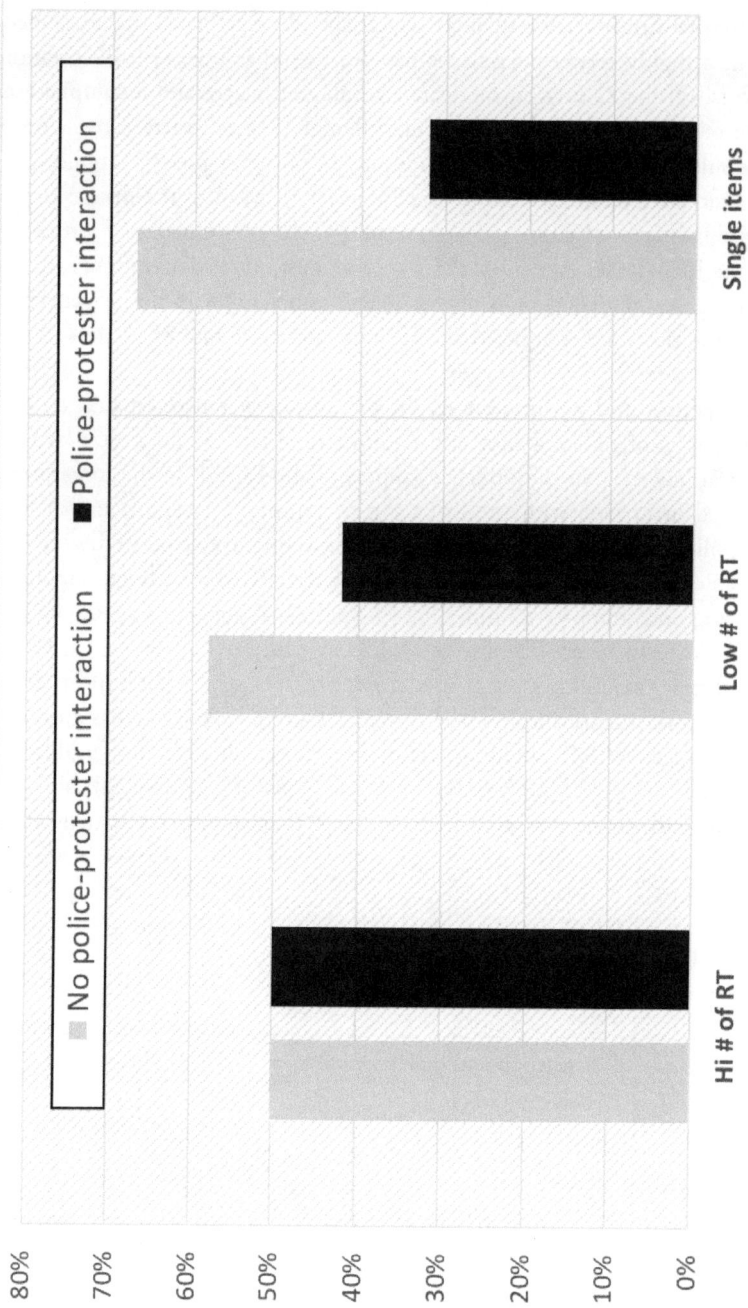

Figure 10.1. Police-protester interaction.

Qualitative findings. We qualitatively analyzed every image that is embedded in a highly retweeted post, which we define as one with 20 or more retweets. Of the 21 images, there are 15 different images and 3 compilations of images. The images fell into four different themes, listed here in order of prominence: police action and equipment, protester presence, police-protester interaction, and solidarity (see Table 10.3). Most of the images depict intense moments of either protesters, the RCMP, or a juncture of protesters and the RCMP (see Appendix A for a catalog of image descriptions). In the following descriptive analysis of the images associated with highly retweeted posts, the posts are numbered in arbitrary order for ease of reference.

Police action and equipment. Eight of the highly retweeted images were photographs of police action and equipment. Post #3 and Post #16 have identical images: two camouflaged snipers of the RCMP in the foreground, crouched in grasses, with the front sniper pointing a weapon, and at least four police in the background in grassland. The image associated with Post #7 depicted one camouflaged member of the RCMP, with gun in hand and weapons strapped to him. He is holding a dog on a leash, who is mid-bark, teeth bared. The image associated with Post #9 shows RCMP in tactical gear (including masks) in a crowded street with what appears to be a police tank/truck/utility vehicle in the background. Post #18 has an image that shows six RCMP standing, facing the viewer, shoulder to shoulder, in the middle of a road surrounded by forest. Tens of other RCMP are scattered behind them. Post #19 has an image that shows two men walking

Table 10.3. Image analysis. Of the 21 images we analyzed, there are 15 distinct images and 3 compilations.

Image type	Theme	Relevant posts
Individual	Police action and equipment	3/16, 7, 9, 18, 19, 20, 21
	Protester presence	1/10/14, 2
	Police-protester interaction	4, 11, 12, 13
	Solidarity	5, 6
Compilation	Police action and equipment	3/16, 8
	Protester presence	8
	Police-protester interaction	15
	Solidarity	17

toward the viewer on a two-way street. Where the road ends, black smoke is billowing up. It is unclear from the image where the smoke originates. However, it is clear from the textual context that the smoke is the RCMP cars burning at the blockade. The image associated with Post #20 is a picture of a camera screen (made clear by "Canon" along the bottom) that shows two-and-a-half RCMP vehicles burning. Flames are visible through the windows and roof and under the vehicle, and black smoke billows from each vehicle. The image posted with Post #21 has a burned RCMP vehicle in the foreground with two people in plainclothes looking at it. One man is looking at the car from several feet back with a red, "no shale gas" sign shaped like a stop sign.

Protester presence. Five of the highly retweeted images primarily depicted protesters. Three posts—Post #1, Post #10, and Post #14—have identical images of protesters, mostly women, standing in the middle of a two-lane road, holding hand drums in the air. Some protesters in the image have flags. The image associated with Post #2 shows six protesters in camouflage and sweatshirts in the foreground. The RCMP are in the background. Post #4 shows protesters sitting and standing around a fire at a camp on the site of the blockade.

Police-protester interaction. Five of the highly retweeted images depicted interactions between police and protesters. Three of those are complete pictures, and two of them are split images. The first of the three whole pictures that show police-protester interaction, Post #11, depicts a woman in a gray hooded sweatshirt kneeling in the foreground with her back to the camera, with her right hand outstretched and holding an eagle feather, which symbolizes "many things including: respect, honour, humbleness, truth, love, natural power, strength, courage, wisdom and Freedom, everything that is positive" to people in the Mi'kmaq First Nation (McIsaac, 2013). Standing several feet in front of her are tens of RCMP.

The second of the three whole pictures, Post #12, shows one RCMP in camouflage and two RCMP in black uniforms. One of the RCMP has a dog on a long leash. Among the RCMP, a person, ostensibly a protester, is facedown on the ground. The last of the whole pictures, Post #13, depicts rows of RCMP standing in profile in the left third of the image; on the right third, four women are standing, with one facing the police line, pointing her left hand; in the middle third, a women has taken a step toward the viewer with a hand drum, looking at the police.

The two split images depict both police action and equipment, and protesters. Post #8 has an associated picture with an inset. The picture is a man in a mask, black T-shirt and red hat, with one arm raised, in front of police cars that are burning. The inset is a man of similar build in the same T-shirt. There is text layered on the image that says, "His name is [redacted]. He is a CIA and RCMP informant. He was responsible for burning RCMP cruisers in an effort to make protesters and Mi'kmaq look bad and to give RCMP a reason for excessive force," and, in a different font, "Was this known informant paid to torch RCMP cars at Elsipogtog? If so, Why?"

Post #15 is a split image of two images that each separately is associated with highly retweeted posts. On the left is the image posted with Post #1/#10/#14, of Mi'kmaq women with hand drums and flags. On the right is the image posted with Post #3/#16 of RCMP snipers in the grass. The images are situated so that it appears that the RCMP are pointing their weapons at the Mi'kmaq women. In the bottom right corner of the right picture, it reads: "Canada's Economic ACTION PLAN," with a black maple leaf.

Solidarity. Three posts had images that expressed solidarity with the protesters. The image associated with Post #5 depicts a woman at a protest holding a sign that says, "Would we let terrorists poison our water supply if they said it created jobs?" This protester does not appear to be at the Elsipogtog's fracking blockade. The image associated with Post #6 shows six children with smiling faces and of varying ages holding signs that together read: "Our parents got arrested for protectin' (*sic*) our water//Say no to fracking//#elsipogtog."

The final image, associated with Post #17, is a compilation of three images of protesters holding signs in the middle of a highway. The signs say "In solidarity with Elsipogtog," "No shale gas drilling; no fracking," and "Don't frack with our water."

Through this analysis, we add nuance to scholarly understanding of social media discourse around this environmental justice movement. In the context of the Elsipogtog First Nation fracking blockade of 2013, peaceful protests are a dominant theme of highly retweeted posts. Heightening the *peaceful* aspect of protests might be a way that the movement is affirmed as within reasonable bounds of citizen action. Additionally, solidarity tends to be textually described more often than visually depicted, whereas police use of force/surveillance is visually depicted more often than textually described. Among the most highly retweeted posts in our qualitative image analysis, nine posts (23%) had eight distinct embedded images depicting police action and equipment. These widely circulated graphic depictions of violence and threatening figures illustrate the most prevalent theme of highly retweeted posts. This visual representation of violence shows tensions between institutional power and marginalized peoples, supporting a narrative of the First Nations Elsipogtog protesters that they are not safe in the physical space in which they are exercising their opposition to shale development, thus encouraging them to turn to hybrid digital spaces to express this dissent. Seven posts (18%) had five distinct images depicting protester presence, including images of protesters at their roadside camp as well as in the road in large numbers, holding hand drums. Five posts (13%) had three distinct pictures and two compilations showing police-protester interaction. Finally, three posts (8%) had embedded images that expressed solidarity with the protesters.

Discussion

A "productive dialogue" between stakeholders with varying backgrounds of "technical expertise" and "local knowledge" (Depoe & Delicath, 2004, p. 10) requires trust, seeking understanding, respect, listening, and equality, which are elements often missing in interactions between Indigenous communities, governmental, and industry stakeholders. The presence of Elsipogtog First Nation protesters at the shale gas exploration site indicates a frustration with the existing system's affordances for addressing conflicts between the environmental priorities of Indigenous peoples with those of government and industry representatives. The conventional system of public participation—as illustrated classically by Arnstein's (1969) original and Connor's (1988) modified ladders—restricts marginalized, dissenting voices. For example, the discursive management that occurs in the solicitation of public comments facilitates bolstering concordant viewpoints while burying or marginalizing critical voices (Carvalho et al., 2016). By using public social media platforms, the protesters pressure stakeholders to react in a public, mediated real-time sphere. They also are taking ownership over a narrative that they may feel is out of their control in the mainstream participatory avenues. Remediating images and visuals as part of the Elsipogtog First Nation protesters' dissenting narrative on social media likely aids in the viral potential of messages (Cox & Pezzullo, 2016).

Our findings here dovetail with existing research on the Idle No More movement on social media in terms of the research questions we ask and the methodology we employ (Callison & Hermida, 2015). We also provide an analysis of this unconventional, uninvited disruptive form of public participation that as of yet remains outside the bounds of public participation scholarship. As social media make space for marginalized dissenting voices—as they enable public attention, heighten visual imagery, and enable engagement by providing a platform (Cox & Pezzullo, 2016)— the traditionally conceived notion of public participation *must* expand to accommodate this space.

Bringing disruptive public participation more formally into the public participation framework is not just a semantic exercise. In the traditional parameters of public participation, there is no clear mechanism for dissenting stakeholders' voices to be meaningfully included in decision-making processes. Social media amplification of resistance efforts is one mechanism that can be used by stakeholders with dissenting voices outside existing channels for formal public participation in decision making on environmental issues. Increasing public attention may put pressure on stakeholders in opposition with those voices, particularly if they have a brand (or office) with a reputation to protect. In the context of the Elsipogtog case study, the themes of the most highly circulated images included police action and equipment and peaceful protests, together contributing to a narrative of a disproportionate reaction of authorities to a peacefully gathered resistance. Though disruptive public participation does not fit cleanly in Arnstein's ladder of participation, its inclusion is integral toward offering an authentic framework by which public participation is occurring in the digital world. Platforms like Twitter evolve over time, and how people use media changes. For example, in 2009, Twitter began facilitating retweets, making it easier for users to repost another user's original content (Liu, Kliman-Silver, & Mislove, 2014). Before 2009, retweets occurred, but users did them manually. Additionally, a series of Twitter updates in late 2013 included facilitating replies and retweets, making them both more streamlined (Sippey, 2013). Research by Liu and colleagues (2014) shows a steady increase in software-facilitated retweets since 2009, with about 20% of total posts in 2013 being retweets. If the trend continued in its current trajectory, then currently it is likely that at *least* 30% of tweets are retweets.

The Twitter of 2013, when the protests whose social media presence we are examining occurred, was a more textual than visual platform. Indeed, on October 29, 2013, just after the Elsipogtog protests, Twitter announced that ". . . starting today, timelines on Twitter will be more visual and more engaging: previews of Twitter photos and videos from Vine will be front and center in Tweets. To see more of the photo or play the video, just tap" (Sippey, 2013, para. 2). Thanks to these changes, Twitter is currently even more of a visual social media platform than the Twitter of 2013. The implications of our research considering the role of images in the social media presence of an environmental justice movement are heightened by this change in how the platform is used.

The recent shift of Twitter to a more visual medium may mean that this platform for disruptive public participation is enhanced compared to its 2013 iteration. As a reminder, by *disruptive public participation*, we mean the involvement of constituents who are from historically marginalized groups or who see themselves as outsiders to decision-making processes and who use digital and social media applications to amplify and document dissent when traditional modes of public participation have been exhausted or are perceived to be ineffective to achieve the goals of stakeholders. In our analysis, certain themes are less often visually depicted than others. For example, in our analysis, "solidarity" was primarily a text-based sentiment, while "police use of force/surveillance and monitoring" was more often expressed in association with an image. Also, posts relaying messages about police action were more often associated with images than were posts with no mention of police action.

If this pattern holds with the increased popularity of visual messaging on the platform, then it is likely that, given similar circumstances, the circulation of these types of police-related images would be increased. Conversely, another possibility, given that visuals are more common now than they were in 2013, is that any heightened effect of having an image embedded in a post may be diluted. Capturing posts from the Elsipogtog movement at the point when Twitter was transitioning into a more visually based platform offers valuable insight into how social media platforms are used in the context of an environmental justice conflict. Future research could apply this analytical framework to the Standing Rock Sioux Tribe's opposition to the completion of the Dakota Access Pipeline in the fall of 2016.

Conclusions

For public participation practitioners, understanding how dissonance manifests on social media is essential. As activists continue to turn to social media to amplify protest, environmental decision makers would do well to look to social media to inform their perspectives on environmental issues. Our analyses contribute an understanding of what types of images and messages are most likely to be amplified and the widest reaching. This research also contributes to understanding polarization on fracking and perceptions of police repression of a marginalized population, as police action and violence are depicted and amplified differently than peaceful protests and solidarity with the protesters.

It is problematic to consider certain groups of people and certain constituents as outsiders to decision-making processes on extractive industries. As we have shown through this case study, social media channels are amplifying voices that are historically left out of environmental decision-making process and public participation processes. In this chapter, we have sought to broaden public participation literature in two ways: first, to include historically marginalized groups of people and their perspectives, and second, to expand the more formalized notions of public participation processes to include the remediation and amplification of voices through digital and social media channels. As our case study shows, repression galvanizes public reaction, which is largely expressed through social media posting. To some extent, the impact of social media on stopping development is unquantifiable. However, an important aspect of examining the impact social media might have is to look at how extensively recirculated impactful posts—those with visual imagery—are. As we argue, the amplification and remediation of Twitter posts about the anti-shale Elsipogtog demonstrations and RCMP crackdown function as a process of *disruptive public participation* by constituents who have been historically marginalized in decision making over natural resource extraction.

Appendix A

What follows is a textual description of every image embedded in a highly retweeted post, along with the associated text of the tweet and the codes that were applied to the image. The post numbers correspond to the numbers used in qualitative image analysis. Twitter handles, or usernames of Twitter users as indicated by an "@" symbol, are redacted for privacy purposes. Grammatical errors in the original tweet text are retained. For access to the images themselves, please contact the corresponding author of this study.

#	Image code	Image description
1	Protesters	Protesters, holding flags and hand drums one handedly over their heads, stand in the middle of the road, facing the view. Six women are visually at the center of the group, and a small child stands in front of them with crossed arms. Trees and a gray sky are visible behind them. *Images #10 and #14 share this image description. Half of Image #15 shares this image description.*
2	Protesters	Six supporters in camoflauge and sweatshirt stand conversing in the foreground. Two police cars are parked along the road in the background, facing to the right of the image. Three police officers stand among the police cars.
3	Police action and equipment	Two snipers are in foreground, crouched in grasses. Front sniper is pointing weapon. There are at least 4 police in background in grassland. *Image #6 shares this image description*
4	Protesters	Protesters convening at a campsite. Six protesters sit in chairs, and three stand. Seven protesters total sit and stand around a fire in the background of the images. Wood is stacked under a tarp in background. One protester sits in the foreground looking directly at the camera.
5	Solidarity	Woman in sunglasses, hood, heavy jacket and neckscarf stands in the middle of a road, looks at camera, and holds sign that says: "Would we let terrorists poison our water supply if they said it created jobs?" People are in the background. The sky is cloudy and bright. The location is unclear.
6	Solidarity	Six children with smiling faces and of varying ages hold signs with between two and four words that together read, "Our parents got arrested for protectin' our water," "Say no to fracking" and "#elsipogtog"
7	Police action and equipment	One members of RCMP is standing akimbo, save one arm slightly outstretched, with weapon in hand in camoflauge clothes, and additional weapons strapped to him. He is holding a dog on a leash, mid-bark with teeth bared. The policeman has a straight face on, and is looking directly into the camera. The policeman and dog are standing in grasses with trees in background.

continued on next page

#	Image code	Image description
8	Police-protester interaction	The main picture has an inset. The picture is a man in a mask, black tee-shirt and red hat, with one arm raised, in front of police cars that are burning. The inset is a man of similar build in the same teeshirt. The text overlaid on the entire image reads, "His name is Harrison Frieson. He is a CIA and RCMP informant. He was responsible for burning RCMP cruisers in an effort to make protesters and Mi'kmaq look bad and to give RCMP a reason for excessive force." (different font): "Was this known informant paid to torch RCMP cars at Elsipogtog? If so, Why?"
9	Police action and equipment	Many police stand in crowded street. All people clearly visible are police in riot gear with masks. What appears to be a stationary police tank/truck/utility vehicle is in the background.
10	Protesters	*See image description for as Post #1.*
11	Police-protester interaction	Woman in gray hooded sweatshirt kneels in foreground with her back to the camera. Her right hand is outstretched and holding what appears to be an eagle feather. Standing several feet in front of her, facing her, are tens of RCMP.
12	Police-protester interaction	One RCMP is dressed in a camoflauged uniform. Two RCMP are dressed in black uniforms. One of the RCMP has a dog on a leash, and the leash is extended. Among the RCMP, a person, ostensibly a protester is facedown on the ground. The ground is dirt, with some grasses and puddles.
13	Police-protester interaction	The profiles of one full row and a partial row of tens of RCMP stand in the left of the image. On the right, four women stand, with one facing the police line, pointing her left hand. In the middle, a woman with a hand drum takes a step toward the viewer looking toward the police on the left.
14	Protesters	*See image description for as Post #1.*
15	Police-protester interaction	Split image: *Image #1* (in short, protesters holding hand drums and flags, with women and a child in front) is on the left. *Image #3* (in short, RCMP snipers pointing weapons toward left of image) on the right. In the bottom right corner of the right picture, it reads: "Canada's Economic ACTION PLAN," with a black maple leaf.

#	Image code	Image description
16	Police action and equipment	*See image description for as Post #3.*
17	Solidarity	Three images of protesters hold signs while standing in the middle of a highway.
		Image A: Three people hold signs. The first two signs are readable, and they say, "in solidarity with Elsipogtog," and "No shale gas drilling; no fracking." Several people are in background.
		Image B: Protesters stand by large truck. A long line of protesters are visible. Most visible is a woman with a sign: "Don't frack with our water."
		Image C: This is a wide photo of protesters scattered on a highway. The front of tractor-trailer is visible. The protesters hold signs that the viewer is unable to decipher.
18	Police action and equipment	Six RCMP stand, facing viewer, shoulder to shoulder, in the middle of a road surrounded by forest. Tens of other RCMP are scattered behind them.
19	Police action and equipment	Two men walk toward viewer on a two-way street. Where the road ends, black smoke billows up.
20	Police action and equipment	A camera screen (made clear by "Canon" along the bottom) frames the picture. Two and a half RCMP vehicles burn. Flames burn through the windows and roof and under the vehicle. Black smoke billows from each vehicle.
21	Police action and equipment	A burned car, assumed to be an RCMP vehicle, is in foreground with two people in plainclothes looking at it. One man looks at the car from several feet back with a red, "no shale gas" sign shaped like a stop sign. Many people are in the middle of the street in the background.

Notes

1. See the chapter in this volume by Hunt et al. on "indecorous voice" for related theorizing on alternative forms of public participation that account for the relationship with social movements outside formal participation processes.

2. Because of their conceptual similarities and their conflation in literature, the terms public participation and citizen participation are used interchangeably throughout this chapter.

3. For an analysis of these hashtags, see Hopke and Simis (2017).

References

Arnstein, S. R. (1969). A ladder of citizen participation. *Journal of the American Institute of Planners, 35*(4), 216–224. doi:10.1080/01944366908977225

Callison, C., & Hermida, A. (2015). Dissent and resonance: #IdleNoMore as an emergent middle ground. *Canadian Journal of Communication, 40*(4), 695–716.

Carvalho, A., Pinto-Coelho, Z., & Seixas, E. (2016). Listening to the public—enacting power: Citizen access, standing and influence in public participation discourses. *Journal of Environmental Policy & Planning.* doi:10.1080/15239 08X.2016.1149772

Caven, F. (2013). Being idle no more: The women behind the movement. *Cultural Survival.* Retrieved from https://www.culturalsurvival.org/publications/cultural-survival-quarterly/being-idle-no-more-women-behind-movement

CBC News Staff. (2013, October 17). RCMP, protesters withdraw after shale gas clash in Rexton. *CBC News.* Retrieved from http://www.cbc.ca/news/canada/new-brunswick/rcmp-protesters-withdraw-after-shale-gas-clash-in-rexton-1.2100703

Chilisa, B. (2012). *Indigenous research methodologies.* Thousand Oaks, CA: SAGE.

Chung, E., & Yoon, J. (2013). An analysis of image use in Twitter message. *Journal of Korea Biblia Society for Library and Information Science, 24*(4), 75–90.

Connor, D. (1988). A new ladder of citizen participation. *National Civic Review, 77*(3), 249–257. doi:10.1002/ncr.4100770309

Cox, R., & Pezzullo, P. C. (2016). *Environmental communication and the public sphere.* (4th Ed.). Los Angeles, CA: Sage.

Depoe, S. P., & Delicath, J. W. (2004). Introduction. In S. P. Depoe, J. W. Delicath, & M.-F. A. Elsenbeer (Eds.), *Communication and public participation in environmental decision making* (pp. 1–10). Albany: SUNY Press.

Depoe, S. P., Delicath, J. W., & Elsenbeer, M.-F. A. (2004). *Communication and public participation in environmental decision making.* Albany: State University of New York Press.

Dreher, T., McCallum, K., & Waller, L. (2016). Indigenous voices and mediatized policy-making in the digital age. *Information, Communication & Society, 19*(1), 23–39. doi:10.1080/1369118X.2015.1093534

Earl, J., McKee Hurwitz, H., Mejia Mesinas, A., Tolan, M., & Arlotti, A. (2013). The protest will be tweeted. *Information, Communication & Society, 16*(4), 459–478. doi:10.1080/1369118X.2013.777756

Elsipogtog First Nation. (2013, June 24). Sovereignty Summer Campaign calling for national solidarity actions and Support for Elsipogtog First Nation front line activists—Idle No More and Defenders of the Land put out national call to action! *Idle No More*. Retrieved from http://www.idlenomore.ca/item/241-solidarity-with-elsipogtogs

Gould, R. V. (2003). Why do networks matter? Rationalist and structuralist interpretations. In M. Diani & D. McAdam (Eds.), *Social movements and networks: Relational approaches to collective action* (pp. 233–257). New York: Oxford University Press.

Guertin, C., & Buettner, A. (2014). Introduction: "We are the uninvited." *Convergence: The International Journal of Research into New Media Technologies, 20*(4), 377–386. doi:10.1177/1354856514542074

Hayes, A. F. (2005). *Statistical methods for communication science*. London: Lawrence Erlbaum Associates.

Hopke, J. E. (2015). Hashtagging politics: Transnational anti-fracking movement Twitter practices. *Social Media + Society, 1*(2). doi:10.1177/2056305115605521

Hopke, J. E. (2016). Translocal anti-fracking activism: An exploration of network structure and tie content. *Environmental Communication, 10*(3), 1–15. doi:10.1080/17524032.2016.1147474

Hopke, J. E., & Simis, M. (2017). Discourse over a contested technology on Twitter: A case study of hydraulic fracturing. *Public Understanding of Science, 26*(1), 105–120. doi:10.1177/0963662515607725

Howe, M. (2013). RCMP bring 60 drawn guns, dogs, assault rifles, to serve injunction on the wrong road. *Halifax Media Co-op*. Retrieved from http://halifax.mediacoop.ca/story/rcmp-bring-60-drawn-guns-dogs-assault-rifles-serve/19358

Howe, M. (2015). *Debriefing Elsipogtog: The anatomy of a struggle*. Halifax; Winnipeg: Fernwood Publishing.

John, S. (2015). Idle No More—Indigenous activism and feminism *Theory in Action, 8*(4), 38–54. doi:10.3798/tia.1937-0237.15022

Liu, Y., Kliman-Silver, C., & Mislove, A. (2014). *The Tweets they are a-changin': Evolution of Twitter users and behavior*. Paper presented at the 8th International AAAI Conference on Weblogs and Social Media, Ann Arbor, MI. http://www.ccs.neu.edu/home/amislove/publications/Profiles-ICWSM.pdf

McIsaac, J. (2013). Significance of the Eagle Feather. *Mi'kmaq Spirit*. Retrieved from http://www.muiniskw.org/pgLegacy06_EagleFeather.htm

McMillan, L. J., Young, J., & Peters, M. (2013). Commentary: The "Idle No More" movement in Eastern Canada. *Canadian Journal of Law and Society / Revue Canadienne Droit et Societe, 28*(3), 429–431. doi:10.1017/cls.2013.47

Mi'kmaq. (2015a). *The Canadian Encyclopedia: Historic Canada*. Retrieved from http://www.thecanadianencyclopedia.ca/en/article/micmac-mikmaq/

Mi'kmaq. (2015b). *First Nations Seeker*. Retrieved from http://www.firstnationsseeker.ca/Micmac.html

Mitchell, B. (2015). Conflict and uncertainty: Context, challenges, and opportunities. In B. Mitchell (Ed.), *Resource and environmental management in Canada* (5th ed., pp. 3–29). Don Mills, Ontario: Oxford University Press.

Mulrennan, M. E. (2015). Aboriginal peoples in relation to resource and environmental management. In B. Mitchell (Ed.), *Resource and environmental management in Canada* (5th ed., pp. 76–84). Don Mills, Ontario: Oxford University Press.

Nabatchi, T., & Leighninger, M. (2015). *Public participation for 21st Century democracy.* San Francisco: Wiley.

Pew Research Center. (2015a). *The Evolving Role of News on Twitter and Facebook.* Retrieved from http://www.journalism.org/files/2015/07/Twitter-and-News-Survey-Report-FINAL2.pdf

Pew Research Center. (2015b). *How do Americans use Twitter for news?* Retrieved from http://www.pewresearch.org/fact-tank/2015/08/19/how-do-americans-use-twitter-for-news/

Saldaña, J. (2009). *The coding manual for qualitative researchers.* Thousand Oaks, CA: Sage Publications.

Sinclair, N. (2014, 2014, Dec. 7). Idle No More: Where is the movement 2 years later? *CBC News.* Retrieved from http://www.cbc.ca/news/aboriginal/idle-no-more-where-is-the-movement-2-years-later-1.2862675

Sippey, M. (2013). Picture this: More visual Tweets. *Twitter News.* Retrieved from https://blog.twitter.com/2013/picture-this-more-visual-tweets

Vowel, C. (2013, 2013, Nov. 14). The often-ignored facts about Elsipogtog. *Toronto Star.* Retrieved from http://www.thestar.com/opinion/commentary/2013/11/14/the_oftenignored_facts_about_elsipogtog.html

Wood, L. J. (2015). Idle No More, Facebook and diffusion. *Social Movements Studies: Journal of Social, Cultural and Political Protest, 0*(0), 1–7. doi:10.10 80/14742837.2015.1037262

Zappavigna, M. (2012). *Discourse of Twitter and social media: How we use language to create affiliation on the web.* London; New York, NY: Continuum International Publishing Group.

Section III

Enacting Horizons of Civic Technology

Chapter 11

Sustainable Stories

Integrated Transmedia as an Ecology of Storymaking

TYLER QUIRING

Transmedia storytelling is the art of world-making.

—Henry Jenkins, 2008, p. 21

Introduction

As Jenkins's line above attests, we need creative ways to build the world we want. Communication research shows that sustainability science provides interdisciplinary spaces for this type of activity (Lindenfeld et al., 2012; Sprain & Timpson, 2012). Yet there persists a need to develop accessible and engaging tools that more fully involve the public in this process. Public participation can be thought of as a family of methods for increasing democratic involvement in complex decision-making processes (Rowe & Frewer, 2000). These methods "range from those that elicit input in the form of opinions . . . to those that elicit judgments and decisions from which actual policy might be derived" (p. 7). I draw on this definition to argue that we must recognize the ways in which public actions continually shape the world to productively engage with these actions through science and policy. As Jenkins (2008) and others posit, this can be thought of as a process called "world-making" (p. 21), which I take to be the means by which communication shapes possibilities for action and reaction.

Transmedia is emerging as a key method to promote engagement through communication. As a form of public participation, transmedia is not merely a tool to increase public understanding and acceptance of science and policy, but also a means to actively shape science and policy with the public in mind and in hand. As I will demonstrate, integrated transmedia provides unique and to date unrealized potential for linking journalism with critical performance to address complex sustainability issues through collaboration between diverse public participants. For the purposes of this chapter, I define collaboration as a process of co-laboring undertaken by groups as diverse as sustainability scientists, policy makers, citizen scientists and decision makers, and media producers as they strive together to create new futures. As new forms of media production such as transmedia story-telling continue to emerge, they provide increased capacity for engagement among environments, people, and stories.

In this chapter, I respond to the urgent need for developing new ways of approaching practical and innovative forms of environmental communication that foreground an ethical commitment to the needs of the world (Cox, 2007). In presenting this response, I explore how—by supporting communication about, for, and with environments—integrated transmedia might enhance public participation in sustainability science. I first outline the development of transmedia as a storytelling phenomenon. I then describe innovative perspectives on transmedia storytelling that highlight its growing capacity. I find that the literature focuses on three distinct uses of integrated transmedia storytelling that relate to its utility for environmental communication, including narrative and documentary, critical performance, and public participation. Next, I offer illustrative examples of two landmark applications of this method that exemplify transmedia as creative journalism and critical performance, respectively. I propose that together, these complementary approaches can enable genres of public participation in which the core commitments of sustainability science deeply intersect (namely, the challenge of addressing how human changes to the planet can be balanced with the need for sustainability and social justice). I then present transmedia storymaking as a uniquely participatory method of environmental communication through an applied case study of my transmedia website focused on the efforts of the New England Sustainability Consortium. Finally, in my discussion I provide a synthesis of insights from my analyses and make a case for transmedia as a reservoir of untapped potential in publicity and engagement efforts related to sustainability science. Specifically, I argue that through an ecology of storymaking, the creative and critical dimensions of integrated transmedia support a distributed politics of engagement uniquely situated to meet the diverse goals and needs of those involved in making worlds through environmental communication.

Transmedia Storytelling: An Emergent Phenomenon

Media scholar Henry Jenkins (2003) is credited with spurring scholarly interest in transmedia, which he described as a media environment where separate story platforms and pieces serve distinct but complementary roles, and each fulfills its unique capacity. He mostly focused on storytelling pragmatics, such as how to consistently entertain an audience by telling a story multiple times across a range of media channels,[1] but also spoke to issues of public participation in his focus on media audiences that modify the texts they consume (2003/2012).[2] While Jenkins was not the first to use the term *transmedia*—it started appearing in compound phrases in the early 1990s (Elwell, 2014; Wheedon et al., 2014)—it was popularized in academic discourse as "transmedia storytelling" when he used the phrase in a column for *MIT Technology Review* (Jenkins, 2003). Transmedia storytelling has since enjoyed extended discussion in media scholarship as a way to describe communication products that employ multiple forms of media in a cooperative instead of merely convergent or competitive capacity. Jenkins (2004) later developed the concept further and described transmedia storytelling as an example of robust convergence that allows an active engagement between audiences and media content across a range of platforms.

When Jenkins again revisited the concept (2008), he provided his most often cited definition: "In the ideal form of transmedia storytelling, each medium does what it does best—so that a story might be introduced in a film, expanded through television, novels, and comics; its world might be explored through game play or experienced as an amusement park attraction" (p. 96). Here, transmedia is distinguished from the more familiar concept of multimedia, which describes a superimposition of forms that are fixed into established molds (a video on DVD, for example). By contrast, transmedia relies on coproductivity and fractal unfolding as key elements of its progressive form. Within such a form, "interactivity" becomes a process of dynamic reconfiguration between the text and the reader. However, Jenkins notes that "relatively few, if any, franchises achieve the full aesthetic potential of transmedia storytelling—yet" (p. 97).

As other scholars began to extend the concept of transmedia story-telling and apply it elsewhere, Jenkins revisited the topic yet again (Jenkins, 2010), citing a growing range of opinions and definitions. Here, he noted an emphasis on entertainment that had defined the discussion up to that time and grappled with broadening the horizons of transmedia to make it more inclusive: "Whatever we call it, transmedia entertainment is increasingly prominent in our conversations about how media operate in a digital era" (p. 944). Understanding the limitations of conceptualizing transmedia as entertainment, he also stated a need to "pull back and lay out some core principles that might shape our development or analysis of transmedia nar-rative and to revise some of our earlier formulations of the topic" (p. 944).

After Jenkins's thinking on the subject became more nuanced and he conceptualized transmedia for purposes beyond entertainment, other scholars began to articulate new uses for the term as well, including narrative and creative approaches to journalism, critical performance, and public partici-pation and collaboration. These uses connect with and extend the purposes of environmental communication and provide ways of addressing complex sustainability issues.

Narrative and Documentary Transmedia

Although transmedia is often focused on narrative fiction, the method has also been used in nonfiction storytelling, which itself has been adapted to research contexts. For example, Nicholas Riggs (2014) writes of his experience reading the autoethnographic thanablog (an online collection of periodic writings about death) of the late Bud Goodall's battle with cancer. Goodall was a communication scholar specializing in the study of storytelling, performance, and narrative, and he charted his experience with cancer online for the public to access. In turn, Riggs uses transmedia as an autoethnographic research method as he folds his own thoughts into his analysis of Goodall's blog. He surmises that "such technology opens a gestural space for public dialogue where there has traditionally been very little surrounding terminal illness and the end of life" (p. 87). This robust conception of human-media relations strongly suggests that transmedia can modify one's sense of self, aligning with Elwell's (2014) explanation that, in an age of digital communication, what we share about ourselves online "is constitutive of the deepest levels of our interiority" (p. 237). The implication here is that transmedia helps activate complex processes of simultaneous individuation and cultural identification.

Taking a closer look at coproduction techniques in the context of journalism, Andreas Veglis (2012) compares transmedia and "cross media" reporting. Veglis describes cross media reporting as a form of parallel journalism, where content is recycled for multiple uses across platforms. By contrast, transmedia approaches take advantage of each medium's unique capacities, as "transmedia is not just about multiple stories, but about creating a rich in-between space, an archive of shared meaning in-between different parts of the story. By using different media, it attempts to create 'entry points' through which users can become immersed in a story world" (p. 315). What this means for journalism specifically is that many different pieces of media can come together through online news platforms to enhance the depth and quality of journalism. Veglis foresees this producing complex forms of textual interactivity that provide many ways of entering a story, allowing for an increasingly engaged and interested readership (p. 322). Similarly, Siobhan O'Flynn (2012) speaks to the engaging qualities of documentary work and describes transmedia as one component of documentary storytelling's "metamorphic" form. Seeking to push the boundaries of "digital narrative," O'Flynn finds that documentary transmedia further emphasizes the presence of local voice and community-centric participation.

Others interested in the boundaries of digital narrative focus on participation as an internal function of documentary approaches to transmedia. For example, Aufderheide (2015) discusses how transmedia commonly takes the form of interactive documentaries delivered through the Web. Specifically, Aufderheide focuses on transmedia projects that deal with complex energy issues such as climate change. While many of her examples provide audiences an enhanced role in the story, these roles tend to relegate participation to the act of gleaning information (p. 77). It is important to note that because Aufderheide is specifically interested in "interactive documentaries," her examples predominantly conform to linear cinematic storytelling norms. Understandably, her analysis of the participatory potential in these transmedia examples focuses on how navigational design presents users with "choice" between limited options within the story. As a result, she mostly avoids an exploration of potential opportunities to have the production of the story itself involve participants.

By contrast, Tamar Ashuri (2012) describes transmedia as a process of social reordering. For Ashuri, activism has become an important niche function of journalism, and transmedia storytelling supports this function by amplifying reader choice and voice. In turn, Cheong & Gong (2010) link civic journalism to transmedia through a discussion of collective intelligence. They describe bottom-up, citizen-driven creation of transmedia texts as a form of cyber vigilantism that "involves the transmediation of information across multiple media platforms as online participants engage with media and with each other to spread information and/or to create new texts" (p. 474). Their key example is "human flesh search, a literal translation from the Chinese nomenclature *renrou sousuo*, [which] involves mediated search processes whereby online participants collectively find demographic and geographic information about deviant individuals, often with the shared intention to expose, shame, and punish them to reinstate legal justice or public morality" (p. 472, emphasis in original). Thus, transmedia platforms in China allow members of the public to circumvent beleaguered systems of conventional justice. In this sense, transmedia itself serves as a way of constituting collective authority, as "the spread of alternative media stories helped subvert and challenge government discourse" (p. 482).

In addition to espousing political resistance, some have used transmedia storytelling to resist the very distinction between fiction and nonfiction. Perhaps most notably, Suleman (2014) dealt directly with this distinction in his analysis of *Four Broken Hearts* (*FBH*), a transmedia storyworld that he produced. As Suleman describes it:

> The FBH experience begins with the announcement that the four main characters, Aaron, Lisa, Zayn and Maha, are meeting for dinner in New York and that "sparks will fly." It then invites the audience to the dinner and asks them to get reservations. This acts as a hook intended to arouse the curiosity of the audience, so that they will want to find out about these characters and why their dinner is an event worth attending. (p. 233)

FBH is an example of integrated transmedia (where the individual pieces work together but do not stand alone) with the primary objective of blurring the boundary between fiction and reality (p. 232). Suleman describes the project as an immersive drama that combines documentary film, additional social media through popular blogging and image-sharing sites, location-based experiences, and live performance, all provided through the project website (http://fourbrokenhearts.com). Along the way, the audience becomes another character in the story, and the main characters emerge as real people whose experiences are documented. Indeed, their social media accounts are actually those of the cast members, and one of the short films found at the end of the website encourages the audience to visit key locations in New York City that mirror the events in the story, as "a chance encounter [with the characters] is always possible" (Suleman, n.d.). Through both its production and presentation, FBH exemplifies how integrated transmedia can operate as both online documentary and critical performance.

Transmedia Storytelling as Critical Performance

As Suleman (2014) and others called into question the boundaries between real and virtual, temporal and spatial, and personal and political (p. 228), some scholars have approached transmedia as a means to problematize other cultural binaries. Chen and Olivares (2014) explore how transmedia "focuses on subversive uses and conceptualizations of media by and for transgender and gender-defiant people in the transnational 'post-digital' age" (p. 245). Here, the conventional roles of media, gender, and nationality are collectively challenged through a common prefix—"trans"—that questions their apparent rigidity. Thus, as it applies to this chapter, transmedia describes a queering of media, or a realization that media can dwell in an "in-between space," to use Veglis's phrase. This approach, in addition to blurring the distinction between producer and consumer, opens a wider capacity for storymaking. It grants that complex issues (such as those studied in sustainability science) need to be explored in complex ways, both in a technical storytelling sense and in a cultural world-making sense.

In a similar spirit, some scholars queer the media through their analysis. For example, in a study of fandom surrounding the television series *Glee*, Marwick, Gray, and Ananny (2014) find that as transmedia texts proliferate, "engagement may provide more than entertainment" (p. 643), as fans find daily opportunities to participate in a robust renegotiation of the politics that mark minority representation in traditional media. In an applied case, Hayles et al. (2014) undertake a critical performance of an alternate reality game called *Speculation*. As they incrementally established a unique character over the course of a year, Hayles et al. found ways of transforming their media experiences through critical play to reinforce an ethos of transmedia existence, embracing the uncertainties of the contingent world it evokes. In these examples, politics extend through personal interaction with transmedia texts and contribute to broader engagement with social justice issues. As

Pezzullo (2003) demonstrates, such issues are entwined with environmental and political processes. Thus, in widening the possibilities for everyday engagement in social justice, transmedia also provides a framework that can complement and extend public participation for and about the environment.

In addition to the above examples of queering, transmedia can blur the lines between producer and consumer to promote engagements with media texts (Jansson, 2013). As Kozel (2007) explains, this is a function of "techne," or the "doing" of communication:

> Fundamentally, techne is about revealing what was concealed, rather than manufacturing or simple instrumentality: techne is a bringing-forth, and technology is a mode of this revealing. In other words, techne is the broad human activity of bringing things into being, while technologies are a modality, or a specific set of practices, within this wider domain. (p. 74)

In relation to transmedia, techne describes acts of storytelling. As the line between producer and audience blurs, stories themselves become the center of focus, storytelling is the "wider domain" within which transmedia operates, and transmedia technologies are the specific modalities through which makers bring stories into being. The key distinction here is that transmedia as a technological mode can present storytelling more richly than multimedia and thus provide opportunities for everyday critical engagement in distributed stories. It can do this because, as discussed earlier, transmedia requires collaborative work as its basic storytelling "techne." If this techne is both critical and constitutive in the ways that much of the work above describes, it follows that the method can complement boundary-spanning efforts in sustainability science and provide environmental communicators a powerful modality through which to make stories together with a broad spectrum of collaborators.

Public Participation through Transmedia Storytelling

For environmental communication to fittingly meet the needs of sustainability science, it must be applied in a manner that embraces and encourages public participation. In fact, Lindenfeld et al. (2012) provide a careful reminder that "sustainability science requires unprecedented levels of collaboration across disciplines and with stakeholders" (p. 31). While these collaborative requirements can appear daunting, online performance itself can be collaborative (Gray, 2012), and transmedia storytelling may be used to provide platforms where public stakeholders can access and make sense of stories, including those related to sustainability science itself. This is possible because transmedia production hands stories over to the reader instead of conveying them through conventional means that limit their application to dissemination or entertainment. While a producer is traditionally needed to provide the storytelling foundation, transmedia encourages audiences to bring their own ideas and experiences to a story, as well as their expectations for their degree of involvement. Indeed, in many ways, the reader is the crux of the transmedia experience.

As we have seen, many who write about transmedia storytelling appreciate the ways it affords readers a deeper level of interaction with stories than is possible with conventional multimedia. Spiridon (2014) went further and argued that transmedia "has a significant impact on the balance between story and discourse, in both directions, and also results in an unprecedented empowerment of the receiver" (p. 33). As opposed to common conceptions of storytelling as "an inexistent narrative activity which is able to convey a story directly to the public, by-passing discourse and its complex sense-making abilities" (p. 35), for Spiridon. transmedia storytelling is more a function of story*making* shaped by the discursive sense-making abilities of an engaged, empowered reader. Invoking a similar ethics, Edmond (2015) found that radio transmedia tends to be more intimate, personal, and localized than mass transmedia entertainment campaigns. For Edmond, common threads in these productions are their emphasis on authenticity of voice and a heightened sense of collaboration between audiences and producers. For example, the radio series *Detroit's Food Economy* featured the voices of Detroit residents who lived in areas where fresh food was difficult to find and afford. Other series have solicited user interaction to guide journalistic investigation or even crowdsource the stories themselves.

One claim particularly amenable to a transmedia ethos is that communication can span challenging boundaries in sustainability science. McGreavy et al. (2013) stipulate that the field of communication should inform sustainability science work that occurs along and across boundaries between scholars, disciplines, and institutions. Furthermore, they argue that communication has a clear role to play in attuning participants to the deep complexity of sustainability science (p. 4215). On the other hand, "one must pay attention to the boundaries (re)created through communication practices" (p. 4205). The implication for environmental communication scholars and practitioners contributing to sustainability science efforts is that these boundaries can measurably shape emerging processes and results. Furthermore, as environmental communicators undertake important work in partnership with stakeholders and then share the results of that work, doing so in a participatory and engaging manner can become a challenge.

While sharing research through media production, for example, may measurably add new dimensions of challenges through the boundaries it (re)creates, it also provides a range of means to meet the unique needs of boundary work in sustainability science. One such means is transmedia storytelling and its aim to productively put multiple forms of media, producers, and consumers in robust interaction with one another. When designed to span the boundaries between disciplines, institutions, and individuals, transmedia provides a method for environmental communicators to work through the complexity of sustainability science and produce platforms that people can use to collaboratively make sense of the world.

ILLUSTRATIVE EXAMPLES: SNOW FALL AND WELCOME TO PINE POINT

Sustainability science is but one form of public participation. Sustainability science cannot itself produce the foundations needed for ideal public participation, but it can build on prior foundations to enhance effective democratic participation through the tools and skills that researchers bring to the collaborative table. To make a case for transmedia storytelling as an important method in sustainability science toolboxes, I next describe and analyze two key examples of integrated transmedia—that is, multiple complementary story pieces collected and conveyed through a single platform—that illustrate its applicability for use as a method of communicating about environmental issues. These illustrative examples demonstrate the journalistic and critically performative functions that transmedia storytelling can perform, and do so through a nonfiction documentary approach. The first is a landmark report by the *New York Times* on the events surrounding a deadly avalanche in Washington state in 2012. The second is an interactive online documentary film about the closing of a mine in Pine Point, Northwest Territories, and the national diaspora and residual environmental damage left in its wake. Both stories serve overlapping but distinct functions, and both have unique lessons to offer environmental communicators seeking to employ transmedia storytelling.

SNOW FALL: *TRANSMEDIA STORYTELLING AS NARRATIVE JOURNALISM*

Perhaps the most well-known example of transmedia storytelling adapted for use in narrative journalism is the extensive investigative report *Snow Fall: The Avalanche at Tunnel Creek* (Branch, 2012). The piece is an online story presented primarily through written text and supplemented by short videos, dynamic animated visuals, and photographs. *Snow Fall* delivers an in-depth account of a large avalanche near the Stevens Pass ski area in Washington that claimed the lives of several skiers on a foray into the backcountry on February 19, 2012. What set *Snow Fall* apart from other online reporting of the time was not only its breadth and depth of content—nearly 18,000 words across 6 thematically organized pages—but also the way it was delivered. In *Snow Fall*, nearly every supplementary piece works in synergy with the written material. Videos are short and to the point and feature clips that speak directly and succinctly to the main content of the page. Animated segments—such as three-dimensional maps of the area, a cross-section view of a mountain slope just before an avalanche, and a replicated view of the avalanche itself complete with approximate locations of buried skiers—are included precisely at the points of the text that are hardest to visualize, anticipating audience needs and delivering content that addresses these needs as readers work their way through the story.

Snow Fall is regarded as a prominent digital journalistic work and integrated transmedia project that in turn came to inspire countless similar projects by other news agencies and subsequently altered reporting approaches (Michel & Ladd, 2015). As an important milestone for integrated transmedia storytelling, *Snow Fall* provides clear precedent for the use of transmedia in environmental communication. Indeed, the piece provides an in-depth exploration of the penalties for transgressing boundaries between nature and culture (while ski areas are typically quite safe, this story occurred in the backcountry where snow conditions are not controlled). Although it reinforces a sense of nature as powerful and dangerous, it does so through an innovative approach that allows the reader to relive a story.

After reliving this story, readers may notice a small icon in the upper-right area of the screen where they can visit a comments page devoted to reader responses to the story. As of this writing, the section is closed to future submissions, with 1,155 comments left by readers. This section includes shortcuts to curated high-impact comments displayed with a "NYT Picks" badge. Many of these comments congratulated the *New York Times* for its research and reporting, but others suggested that readers' story experiences constitute *Snow Fall* as a milestone along the way to new kinds of report-ing. For example, one reader wrote, "While 'experiencing' this article, I felt both present at and omniscient about the tragedy. The mixture of first and third person, interviews and simulations, visual and audio, was fantastic. Now I feel like I witnessed the birth of a new form of journalism." Another (claiming 25 years of backcountry skiing experience) described the piece's form as transportive:

> This is more than a perfectly written piece. This is almost a new form of media so unbelievably deep that . . . I felt the powder billowing turns, I felt the sweat at the end of the run in the warmer, lower elevations. I also felt the frantic fear and the the [sic] physical pain from willing, and yearning to find your partner alive. I remembered the taste of adrenalin and blood. Having lost my girlfirend [sic] to an avalanche that also nearly claimed me, this was a difficult piece to read. It is also a piece I will add to my library.

Snow Fall demonstrates the culmination of a tension between multi-media (many media compressed into one form) and transmedia as well as a tension between franchised versus integrated transmedia storytelling. It contains remnants of multimedia approaches to news reporting, relying on text-centric and completionist storytelling, but also productively uses transme-dia storytelling to establish a new paradigm of news reporting. At the same time, while it draws on the familiar features of the *New York Times* as far as the *Times'* format might be called a "franchise" of news reporting, the piece goes beyond this to make a clear case for integrated transmedia as a powerful method of documentary storytelling. Furthermore, it demonstrates that this approach can adequately attune users to environments, as its experiential qualities provide many opportunities for deeper engagement with the text. As the readers above report, these opportunities also provide some level of engagement with the environment from a computer screen. However, these opportunities are not always undertaken as public participation. As can be seen through *Snow Fall*, the *New York Times* pioneered this emergent form of journalism to increase reader awareness and textual engagement. Textual engagement does not necessarily translate to public participation, but as we will see, it can. The next example illustrates another transmedia work that serves as a milestone on the way to using transmedia to enhance public participation. It also demonstrates that collaborative storytelling can produce highly affective experiences for readers.

WELCOME TO PINE POINT: TRANSMEDIA STORYTELLING AS CRITICAL PERFORMANCE

Welcome to Pine Point (Shoebridge & Simons, n.d.) is an interactive online documentary film produced and distributed by the National Film Board of Canada. It discusses the place-based and discursive contexts that constituted the mining town of Pine Point, Northwest Territories, from its establishment in the early 1950s to its closure in the late 1980s. Specifically, it explores the aftermath of the mine's closure, including residents' loss of their homes and the national diaspora this prompted. The documentary brings the Pine Point of the past into being through a range of transmedia techniques, including reminiscent poetry and prose written by the producers, archived photographs and video, original audio interviews, photos and videos collected by the producers and residents, drawings and visual reconstructions of and by Pine Pointers, and an original soundtrack by Canadian band *The Besnard Lakes*. Together, these elements provide a rich space for readers to co-perform the experience of living in Pine Point.

Performative transmedia in *Pine Point* highlight meaning making as a collaborative activity. Herman (2004) states that "at the heart of each narrative . . . is the same problem—namely, how to make sense of a transformation by which a character becomes a member of a different species" (p. 57). This metamorphosis is central to making sense of *Pine Point*. In one climactic scene, we are shown how physically and socially disorienting a diaspora can be. The screen shows two side-by-side camera shots, one of a house in Pine Point during the town's heyday, another of the same location years later, now an empty, houseless lot. As the videos pan simultaneously across the space, sorrowful music crescendos in the background. The effect is swelling emotion, an overwhelming sense of loss, but also a revelation on the resilience of personal attachment to place.

In *Welcome to Pine Point*, the town becomes re-performed through collaborative transmedia by a spectrum of coproducers, including the official producers through their vision in piecing together the story, the town's residents through their memories recalled across many media pieces, and also the reader through his or her unique exploration of the narrative. Through these textual engagements, the piece establishes a form of collaborative storymaking actualized through an integrated transmedia performance. This is powerfully demonstrated as resident experiences are filtered and told through sociospatial contexts, then retold by producers in a myriad of ways, and finally retold yet again by audiences through their own interaction and understanding of the performed text. Together, these storymakers co-create the worlds of both Pine Point as a town that exists only in memory and *Pine Point* as a transmedia artifact.

The illustrative examples above demonstrate how transmedia has been used within journalism and documentary production and show the expanded potential that transmedia affords for textual and environmental engagement as well as collaborative storytelling. As such, they are key milestones for the deployment of transmedia within public participation contexts. Next, I share an analysis of an integrated transmedia website I created that uses collaborative transmedia storymaking to support natural resource management. This applied case study was undertaken as part of an interdisciplinary research project that exemplifies how transmedia can be used to augment public participation approaches within sustainability science.

Applied Case Study: *Safe Beaches, Shellfish, and You*

The New England Sustainability Consortium (NEST) is an interinstitutional, transdisciplinary team of sustainability researchers in New England. The goal of the consortium is to enhance the use of science in decision making in New Hampshire, Maine, and Rhode Island. NEST's first project was funded through a grant provided by the National Science Foundation and focused on science and decision making within the beach and shellfish industries in Maine and New Hampshire. As a NEST researcher at the University of Maine, I became interested in making a story with NEST constituents to complement our public reporting strategies and contribute to our goal of enhancing the use of science in decision making. For these reasons, I co-created an integrated transmedia website with NEST researchers and stakeholders through an iterative project design that could adapt based on lessons that surfaced in the storymaking process.

The approach I took to the transmedia website was informed partly by much of the scholarship I have explored above, but also by important lessons learned from integrated transmedia pieces like *Snow Fall* and *Welcome to Pine Point*. These pieces were valuable and instructive for the breadth and depth of their content, which provided multiple avenues for reader interaction. The high-quality, extensive reporting of *Snow Fall* delivered through a range of interconnected and engaging technologies had demonstrated that in-depth transmedia reporting could create additional spaces for engagement with environments. *Welcome to Pine Point* had shown that documentaries, while usually linear, could be constructed in ways that acknowledge readers' empowered and critical work with narratives that ask difficult questions about the complexities of human interactions within natural resource contexts. Yet I wanted to combine reporting *about* New England's coastal environment with opportunities for readers to critically perform *in* that environment, no matter where they were. Furthermore, I sought to create a space where those interested in NEST could experience its commitment to collaborative research and co-create the story through their interaction and further participation.

The project's eventual form was a website called *Safe Beaches, Shellfish, and You* (*SBS&Y*) housed at http://nest.maine.edu, where I combined written reporting with photography, documentary audiovisual material, interactive graphics, and immersive experiences. The result was a nine-page integrated transmedia storymaking website that was the product of collaboration and that provided further engagement through interactivity, narrative performance, and outreach opportunities. I coded *SBS&Y* in the Bootstrap web development framework, which delivers content through responsive web design that automatically formats content to a range of device types and screen sizes. Importantly, the *SBS&Y* narrative is organized thematically instead of chronologically to allow for multiple ways of making sense with and moving through the stories that become co-performed on this platform.

WEBSITE WALKTHROUGH

When readers visit the landing page (see http://nest.maine.edu/), they are greeted with a welcome screen that labels the site as a storytelling platform, provides some written context on NEST, and invites them to "Explore the Story."

Upon clicking the prominent entry button, the reader comes to the main page of the site, seen below. The top section of this page provides important orienting information, including a banner image of shellfish harvesters working on a beach at sunrise, a series of in-page tabs answering "who, what, how, and why" questions about NEST and the Safe Beaches and Shellfish project, and an interactive timeline that provides an overview of how activities progressed within the project. If the reader scrolls down, he or she finds a section of the page that introduces the concept of sustainability, presented by three featured individuals including a graduate student on NEST, a lobsterman who serves as a tribal chief, and a NEST postdoctoral researcher. Underneath is a short audio piece that weaves together interviews in which each of these individuals provides diverse definitions of sustainability.

The reader will also notice thematically ordered navigation buttons at the top of the page, including "Beaches," "Shellfish," "Collaboration," "Engagement," "Get Involved," and "About" (see http://nest.maine.edu/beachesandshellfish/main).

The content on these pages follows a consistent format. First the reader finds a banner image at the top that provides a visual example of the page's topic. Immediately below is a written narrative that introduces NEST's research and involvement with that topic. Following or embedded within the story are visual, interactive, and/or media pieces that further illustrate the written material and provide a range of opportunities for meaning making. Specifically, the two pages that deal directly with natural resources ("Beaches" and "Shellfish") include additional experiential and immersive media. On the "Beaches" page, there are two unique ways that the reader can be a co-creator of the story. In the middle of the page after the main written piece, readers can find an embedded interactive window. This window includes "photospheres," images of five beaches in New Hampshire and Maine presented through scrollable 360-degree imagery that sports interactive features similar to Google's popular Street View technology. Just below is an audio piece of the sounds at one beach. Below that are panning videos of 4 beach locations.

Together, these materials provide opportunities for the reader to imagine him- or herself at these locations. The purpose is to allow for the co-creation of a lived sense of understanding as much as possible through a transmediated experience. For example, a key piece on the "Shellfish" page is a situated audiovisual account of a clam harvester digging on a mudflat in Maine's Frenchman Bay. To create this account, a collaborator and I outfitted the harvester with a head-mounted GoPro action camera and ear-mounted binaural microphones. Binaural audio is a technique for recording sounds in which microphones are placed in such a way as to emulate the position of human ears and thus capture sounds as one would naturally hear them. Situations recorded in this manner have the potential to convey the unique aspects of a situation much more effectively than traditional video and audio coverage from an external perspective because the recording devices are situated on the subject similarly to how the audience will experience the video as well. The viewer sees the scene in much the same way as the person wearing the camera and (if wearing earphones as he or she listens) hears the same sound that entered the subject's ears as well. In this way, there is a high degree of overlap between the original and secondhand experience as the subject and viewer are brought closely together.

The "Collaboration," "Engagement," and "Get Involved" pages build upon the experiences developed through earlier sections of the website. These pages progress from depicting traditional collaborative research approaches to sharing a model of public participation in line with Rowe and Frewer's (2000) ideal. The first of these pages, "Collaboration," focuses mostly on NEST team interactions. It includes text that describes how the project foregrounded interdisciplinary and intersectionality as key research priorities. The page also includes three videos that focus on events at an all-team NEST meeting. These meetings were opportunities for NEST researchers from all states, institutions, and disciplines to come together and collaboratively share the shaping of the project. The first video on the page consists of a researcher interview combined with footage of team members interacting during planning activities. The other two videos on the page contain short excerpts of team member reactions to the meeting style and objectives.

The "Engagement" page follows in a similar style as the previous page, but expands to describe engagement beyond the NEST research team. This includes an overview of two innovative courses provided across NEST institutions. These courses connected students with concepts informing the Safe Beaches and Shellfish project and provided opportunities for engagement and outreach with the public as well. The page uses a similar format to the "Collaboration" page, with a text write-up that includes descriptive information as well as additional short videos. The first video combines an interview with a NEST team member teaching the courses with footage shot during one class period. For the course's final project, students conducted their own outreach or engagement activity. The second video on the page focused on one student's final project, an engaging lecture about watersheds delivered to a class of middle-school students in Maine. Together, these videos demonstrate a turn from an emphasis on research collaboration to a broader spectrum of engagement opportunities with the public.

Continuing this turn, the final main page of the website is the "Get Involved" page, where readers are invited to participate in sustainability science in a number of ways. The page begins with a written story about the experiences of a public stakeholder that NEST worked with and includes their own suggestion that the public become more involved in sustainability science. The bottom section of the page includes information on a range of academic and public citizen science programs and other participatory initiatives connected with sustainability. These include an invitation to become formally involved as a NEST stakeholder, links to a range of citizen science programs that are run by various NEST institutions and are open to the public, and information about citizen science participatory initiatives that are not directly connected with NEST but complement its focus. For each opportunity, links are provided where the reader can find more information, and at the end of the page readers are also invited to get in touch directly with NEST through the use of a team contact email address where queries can be directed.

Last, the "About" button in the top navigation bar will bring the reader to a section with more information about transmedia storytelling, credits, links to additional resources, and a list of references. Together, the website's pages leverage transmedia techniques to offer a suite of engagement opportunities at varying levels of public participation. These range from participation at the level of website interaction through engaging content, to examples of NEST's public participation focus through sustainability science, to information that provides an impetus for further public participation through direct action in a range of available programs relating to citizen science.

This tiered approach to engagement reflects three strategies for public participation through transmedia that complement NEST's approach to sustainability science outputs: engagement at the level of collaborative team operations connected to conventional outreach activities; engagement at the level of textual interactivity; and engagement as a process of collective participation in emergent opportunities. Next, I discuss how website usage data collected after the website's launch connect with key transmedia functions introduced in the literature and developed through the illustrative examples. I then argue that these connections suggest that integrated transmedia can best enhance public participation in sustainability science when it is approached as an ecology of storymaking.

INTERPRETATION OF WEBSITE USAGE DATA

From the launch of the website on November 4, 2015, website user data were collected over a nine-month period using Google Analytics. The data were cleaned by removing traffic originating from bots.[3] The cleaned data revealed 2,206 page views across 554 website visits from 308 visitors with an average session duration of 3:39 (three minutes and thirty-nine seconds). Additionally, these visits yielded an average of 3.98 pages visited per session and a bounce rate[4] of 33.03%. The ratio of new visitors to returning visitors was 1:1. The average number of pages viewed per session for new visitors was 3.32 versus 4.64 for returning visitors. The average session length for new visitors was 02:23 versus 04:55 for returning visitors.

Website user data indicate a low impact factor in terms of user and session count but a relatively high impact factor in terms of page views and session length. In other words, although the website did not generate a particularly high level of visitation, data suggest that those who did visit were highly engaged by the content. This is reflected in several metrics. In particular, visitors on average viewed nearly half of the site's pages per session (~4 of 9), dividing their time fairly equally across pages visited. Also, the high percentage of returning users (50%) demonstrates that the site brought half of all individual visitors back again, indicating engagement that unfolds over time. Finally, average session length versus pages viewed for new visitors (02:23 and 3.32, respectively) compared to returning visitors (04:55 and 4.64) are important metrics, which may indicate an engagement spread between the public (assumed to be less likely to return) and NEST team members (assumed to be more likely to return).

These data are thus also instructive in suggesting that transmedia engagement is complementary to but not synonymous with public participation. The website metrics do certainly demonstrate that public participation with online storytelling about sustainability science can be enhanced using engaging techniques such as transmedia. However, other dimensions are worth considering. In fact, location data suggest that when undertaken as story*making* (as with *Safe Beaches, Shellfish, and You*), transmedia can serve vital functions for internal collaboration on sustainability science projects. When website visits were expressed in terms of location, a geographical pattern emerged where NEST institution location became one of the most important factors describing website visits. Globally, the United States accounted for 96.57% of traffic to the site. Together, Maine and New Hampshire accounted for 79.44% of all U.S. visits (54.95% and 24.49%, respectively). Within these states, cities with lead NEST institutions accounted for between 46.26% (Orono, Maine) and 78.63% (Durham, New Hampshire) of the regional total.

While these figures do not surpass conventional metrics of online success, I maintain that transmedia powerfully enhance engagement efforts. For example, a NEST PhD student explained to me that when she was preparing her public dissertation defense, she spent hours exploring the transmedia website again to increase her familiarity with the range of research that had been done on the project. NEST faculty have cited the transmedia website on slides in presentations and use it as an interactive "press kit" when reaching out to stakeholders or other members of the public. I have also used the website as an outreach tool when engaging with the public and as a personal resource when referring to project information cataloged there.

Discussion

We have seen extensive variety in both the research and practice of transmedia. We have also seen that the development of integrated transmedia ensures that stories can function as an organizing apparatus while still providing a space that allows for diverse approaches to sense making. How then can integrated transmedia storytelling support the efforts of environmental communicators seeking to employ public participation in sustainability science? Based on the materials provided here, I argue that transmedia can be an evocative and affective enhancement of engagement strategies. Far from the traditional, rigid sense of story*telling* that Spiridon (2014) rejects or the navigational channeling that Aufderheide (2015) warns of, the examples I have provided illustrate that transmedia is likely at its best when producers and readers work together as story*makers* who use the innovative tools of integrated transmedia to innovative ends.

The *New York Times*' immersive 2012 report *Snow Fall* paved the way for future efforts, including efforts that could more fully embrace the role of the reader as participants in the story. *Welcome to Pine Point* further refines this approach, constantly inviting the reader to co-perform the text as he or she participates in making meaning out of a collective trauma. My transmedia project *Safe Beaches, Shellfish, and You* seeks to innovate further, where the processes of both making and reading the story are a process of collaborative coproduction that constitutes a rich ecology of integrated transmedia storymaking. In this approach, not only do story segments cohabitate as distinct but complementary species, but engagement itself becomes distributed among a range of storymakers.

As a media ecosystem, *Safe Beaches, Shellfish, and You* provides a shared platform for parties interested in the work of the New England Sustainability Consortium to orient around sustainability science through collaborative storymaking. Through a process of iterative development informed by collaborators, the site foregrounds engagement at the level of storymaking. Within this ecological approach, readers may engage with the text not just as passive consumers but also as active co-creators whose personal experiences and values come to augment instead of limit the storytelling that is possible. In so doing, the project serves the ethical commitments of environmental communication by forming possibilities for public participation.

While the first NEST project has now come to a close, *SBS&Y* will inform further efforts within the New England Sustainability Consortium. In fact, the first of these is already under way. The Consortium's next project on the Future of Dams is developing a range of tools that can be used in decision making about dams. As both method and toolbox, transmedia is already shaping how we think about interdisciplinary integration and stakeholder engagement. Moving forward, a transmedia ecology of storymaking will provide many opportunities to help stakeholders such as communities, regulators, and scientists negotiate key trade-offs as they make pressing decisions about the role of dams in the future of New England and the United States.

Conclusion

Although transmedia has had an eclectic intellectual and practical history, the diverse theoretical and applied approaches described above reveal that this emergent method for storymaking can provide environmental communicators with a powerful set of tools to enhance public participation. Furthermore, integrated transmedia storymaking allows stories to continually adapt by narratively and experientially involving audiences in a robust manner that includes textual and extra-textual engagement. While there is much work yet to be done to demonstrate the full potential of the form, the examples I have shared provide a compelling case for integrated transmedia storymaking as an innovative approach to working across boundaries in sustainability science and building bridges between scholars, practitioners, and stakeholders. As Jenkins observed, there is an art to this, one that relies on engagement as a key element of the meaning-making process.

The two case examples and single applied case study explored above demonstrate that transmedia approaches are already expanding the scope of available means and modes of public participation and community engagement. As a philosophy of making and telling stories, transmedia acknowledges meaning making as a collective, distributed process. As a media modality, it provides equipment for tracing and sustaining hybrid experiences that become entangled as the line between audience and producer continues to blur. Additionally, these are domains for incorporating place-based cultural and activist practices, diverse forms of science, and inventive technological approaches within a recognizable but adaptive framework. It is this framework that I call integrated transmedia storymaking.

There are constraints here as well, including for data accessibility and preservation. Additionally, transmedia storymaking has not yet reached developmental maturity. However, as described above, its great value lies in demonstrating an ecological orientation to public participation. Within this orientation, ecology is a way of knowing as well as a metaphor for interconnectedness. Because NEST is itself an environment in which researcher and stakeholder participation-as-process strives for sustainability, integrated transmedia became a way of giving the story even more room to grow. In this realm, the ideal function of an integrated transmedia website shifts from a platform for publicity to a habitat within which stories do not just exist, but also thrive. An early goal of my project was to employ transmedia storytelling to share information about sustainability science. The website reached that goal, but along the way it evolved to illustrate the unfilled capacity that transmedia still holds for engagement potential.

We have seen that transmedia storytelling originally developed as a means to describe mass-produced entertainment narratives that served the purposes of elite media corporations. By contrast, the transmedia storymaking I outline encourages a distributed politics of engagement that relies on the free exchange of input among collaborators. Ideally, transmedia storymaking should develop as a highly distributed set of activities that draws on the varied skills of these diverse collaborators, including but not limited to blogging by multiple stakeholder groups, de-localized content collection through social media submission, open authorship, and iterative development cycles that routinely draw on feedback to earlier versions. This field of activities will draw closer to strong public participation as more stakeholders, knowledge types, and storytelling approaches are included. Furthermore, as future integrated transmedia storymaking efforts are undertaken, "producers" would do well to anticipate what data they need to collect along the way to more fully reflect on their project's efficacy and also to inform future efforts.

More to the point, in taking a storymaking approach to engagement, it becomes clear that the next step to achieving robust public participation through transmedia is involving diverse groups of citizens, policy makers, and researchers as official coproducers. Transmedia storymaking in the context of public participation begs for—and indeed provokes—ruptures in the barrier between traditional notions of "producer" and "audience" while foregrounding engagement not as an end point, but as a method for collaboration. For sustainability projects such as NEST, integrated transmedia storymaking means soliciting input and reporting from all corners of a collaboration as early as possible and in a systematic, open, and voracious manner. In this approach, communication itself becomes a catalyst for further extra-digital engagement in a process of iterative, dynamic participation among politically attuned publics.

Acknowledgements

The author thanks Bridie McGreavy for her advice and substantive feedback, Jennifer Moore and Laura Lindenfeld for their mentorship, the coproducers of *Safe Beaches, Shellfish, and You* for their world-making contributions; and the organizers of the 5th Iowa State University Summer Symposium for their hospitality and support. This research was supported by National Science Foundation award #11A-1330691 to Maine EPSCoR at the University of Maine.

Notes

1. Jenkins's key example is *The Matrix*, an entertainment franchise including three blockbuster films and spinoff anime, videogames, and comic books. For Jenkins, this web of texts "places new demands on consumers and depends on the active participation of knowledge communities" (2008, p. 21).

2. While Jenkins does not make the link between entertainment and engagement explicit, in an analysis of climate change films, McGreavy and Lindenfeld (2014) argue that "how we entertain ourselves shapes our understanding of the world and our actions in it" (p. 133).

3. Bots are automated servers that crawl websites, generating nonhuman traffic that can either be useful or nefarious, but in either case dilute metrics of site visitation.

4. Bounce rate describes the percentage of users who exited the site from the same page they entered without substantially interacting with that page.

References

Ashuri, T. (2012). Activist journalism: Using digital technologies and undermining structures. *Communication, Culture & Critique, 5*(1), 38–56.

Aufderheide, P. (69). Interactive documentaries: Navigation and design. *Journal of Film and Video, 67*(3–4), 69–78.

Branch, J. (2012). Snow Fall. *The New York Times*. Retrieved from http://www.nytimes.com/projects/2012/snow-fall/#/?part=tunnel-creek

Chen, J., & Olivares, L. (2014). Transmedia. *TSQ: Transgender Studies Quarterly*, *1*(1–2), 245–248.

Cheong, P. H., & Gong, J. (2010). Cyber vigilantism, transmedia collective intelligence, and civic participation. *Chinese Journal of Communication*, *3*(4), 471–487.

Cox, R. (2007). Nature's "crisis disciplines": Does environmental communication have an ethical duty? *Environmental Communication* *1*(1), 5–20.

Edmond, M. (2015). All platforms considered: Contemporary radio and transmedia engagement. *New Media & Society*, *17*(9), 1566–1582.

Elwell, J. S. (2014). The transmediated self: Life between the digital and the analog. *Convergence: The International Journal of Research into New Media Technologies*, *20*(2), 233–249.

Gray, J. M. (2012). Web 2.0 and collaborative on-line performance. *Text and Performance Quarterly*, *32*(1), 65–72.

Hayles, N. K., Jagoda, Patrick, & LeMieux, P. (2014). Speculation: Financial games and derivative worlding in a transmedia era. *Critical Inquiry*, *40*(3), 220–236.

Herman, D. (2004). Toward a transmedial narratology. In M. L. Ryan (Ed.), *Narrative across media: The languages of storytelling* (pp. 47–75). Lincoln, NE: University of Nebraska Press.

Jansson, A. (2013). Mediatization and social space: Reconstructing mediatization for the transmedia age. *Communication Theory*, *23*(3), 279–296.

Jenkins, H. (2003/2012). Quentin Tarantino's Star Wars? Digital cinema, media convergence, and participatory culture. In M. G. Durham & D. M. Kellner (Eds.), *Media and cultural studies: Key works* (pp. 452–470). West Sussex, UK: John Wiley & Sons.

Jenkins, H. (2004). The cultural logic of media convergence. *International Journal of Cultural Studies*, *7*(1), 33–43.

Jenkins, H. (2008). *Convergence culture: Where old and new media collide* (2nd ed.). New York, NY: New York University Press.

Jenkins, H. (2010). Transmedia storytelling and entertainment: An annotated syllabus. *Continuum: Journal of Media & Cultural Studies*, *24*(6), 943–958.

Kozel, S. (2007). *Closer: performance, technologies, phenomenology*. Cambridge, MA: MIT.

Lindenfeld, L. A., Hall, D. M., McGreavy, B. M, Silka, L., & Hart, D. (2012). Creating a place for environmental communication research in sustainability science. *Environmental Communication*, *6*(1), 23–43.

Marwick, A., Gray, M. L., & Ananny, M. (2014). "Dolphins are just gay sharks": *Glee* and the queer case of transmedia as text and object. *Television & New Media*, *15*(7), 627–647.

McGreavy, B., & Lindenfeld, L. (2014). Entertaining our way to engagement? Climate change films and sustainable development values. *International Journal of Sustainable Development*, *17*(2), 123–136.

McGreavy, B., Hutchins, K., Smith, H., Lindenfeld, L., & Silka, L. (2013). Addressing the complexity of boundary work in sustainability science through communication. *Sustainability, 5*, 4195–4221.

Michel, J. P., & Ladd, M. (2015). "Snow Fall"-ing special collections and archives. *Journal of Web Librarianship, 9*(2–3), 121–131.

O'Flynn, S. (2012). Documentary's metamorphic form: Webdoc, interactive, transmedia, participatory and beyond. *Studies in Documentary Film, 6*(2), 141–157.

Pezzullo, P. C. (2003). Touring "Cancer Alley," Louisiana: Performances of community and memory for environmental justice. *Text and Performance Quarterly, 23*(3), 226–252.

Riggs, N. A. (2014). Leaving cancerland: Following Bud at the end of life. *Storytelling, Self, Society, 10*(1), 78–92.

Rowe, G., & Frewer, L. J. (2000). Public participation methods: A framework for evaluation. *Science, Technology, & Human Values, 25*(1), 3–29.

Shoebridge, P., & Simons, M. (n.d.). *Welcome to Pinepoint.* Retrieved from http://pinepoint.nfb.ca/#/pinepoint

Spiridon, M. (2014). Repairing the paradigm: The fate of the narrative in a "network culture": *Revista Romana de Jurnalism si Comunicare—Romanian Journal of Journalism and Communication, 3*(45), 32–37.

Sprain, L., & Timpson, W. M. (2012). Pedagogy for sustainability science: Case-based approaches for interdisciplinary instruction. *Environmental Communication, 6*(4), 532–550.

Suleman, B. (n.d.). *Four broken hearts.* Retrieved from http://fourbrokenhearts.com

Suleman, M. B. (2014). Like life itself: Blurring the distinction between fiction and reality in the *Four Broken Hearts* transmedia storyworld. *Journal of Media Practice, 15*(3), 228–241.

Veglis, A. (2012). From cross media to transmedia reporting in newspaper articles. *Publishing Research Quarterly, 28*(4), 313–324.

Wheedon, A., Miller, D., Claudio, P. F., Moorhead, D., & Pearce, S. (2014). Crossing media boundaries: Adaptations and new media forms of the book. *Convergence: The International Journal of Research into New Media Technologies, 20*(1), 108–124.

Chapter 12

Eco-Apps and Environmental Public Participation

ELI TYPHINA

Environmental public participation (EPP) initiatives, such as citizen advisory boards, public hearings, and volunteer environmental monitoring, involve multiple stakeholders who affect and are affected by complex environmental problems. By involving multiple people in EPP initiatives, stakeholders present new and alternative perspectives, leading to more diverse solutions than those identified without stakeholder involvement (Jentoft & Chuenpagdee, 2009). However, the structure of traditional EPP initiatives can limit stakeholder involvement. For example, the location or meeting times of events can conflict with stakeholders' work and family responsibilities as well as stakeholders' ability to physically attend events (Smith & Norton, 2013). Additionally, a lack of access to resources, a lack of empathy or concern of all perspectives, and an inability to meaningfully inform decision-making processes can further reduce stakeholder participation (Senecah, 2004).

As a way to allow stakeholders to actively and meaningfully contribute, environmental planners have started using Web 2.0 technologies such as discussion forums. Web 2.0 features used for EPP include commenting, liking, sharing, or purchasing an item to support conservation efforts (Büscher, 2014). These features allow stakeholders to co-create content in an "architecture of participation" instead of passive viewing (O'Reilly, 2005, para. 24), which leads many scholars to argue that Web 2.0 is democratic, empowering, and engaging for the individuals who use it (Castells, 2007; Reynolds, 2006; Shirky, 2008; Tapscott & Williams, 2006). Through co-creation, Web 2.0

can help EPP planners overcome challenges of traditional approaches, such as the one-way transmission of messages at public hearings and the limited involvement of diverse community members on citizen review panels (Depoe, 2004). For example, local governments in Brisbane, Australia, found that use of Web 2.0 platforms, such as Facebook, Twitter, blogs, and online forums, afforded citizens access to more resources and increased discussion when compared with in-person meetings (Fredericks & Foth, 2013).

Furthermore, the affordances of Web 2.0 on mobile phones place the resources and connections needed to solve environmental problems in the hands of stakeholders. Mobile phones offer a degree of mobility, connectivity, and personalization that is not possible on a laptop or desktop computer. Mobile phone applications using Web 2.0, such as social media applications, are well-known for their ability to prompt rapid co-creation of content, especially location-based geotagged content. The original Foursquare application digitally connected friends, places, and information through gamification, features that reorder users' daily lives and changed their relationships with physical places (Schwartz, 2014). If the outcomes of Foursquare are possible in relation to environmental issues, stakeholders and experts might gain new perspectives and form new relationships through use of apps.

Additionally, mobile applications aiming to engage people in environmental public participation, which I call eco-apps, offer a unique alternative to traditional EPP initiatives because users can upload locative content, take photos with the mobile's camera, and offer real-time feedback. Citizen science applications, such as Wildlab Bird and ProjectNoah (discussed in this chapter), offer app users a way to collect scientific data and gain deeper understanding about the ecosystems around them. Similarly, the MyActions and Sydney photo walk eco-apps (also discussed in this chapter) afford users ways to connect with community members and collectively explore environmental problems and solutions. Thus, eco-apps offer the chance for stakeholders to gain technical competency and "fluency in the language(s) of science" required for genuine participation in EPP initiatives (Kinsella, 2004, p. 85). Additionally, eco-apps offer mechanisms for *civic discovery* by enabling users to collaboratively learn about and discuss the environmental situation of concern (Depoe, 2004; Walker et al., 1996). Even with such potential, very little research-based guidance exists to help EPP planners effectively design eco-apps to support stakeholder involvement.

Therefore, in this chapter I analyze the structures and features of 11 eco-apps with Web 2.0 features and the user outcomes of these apps.

I focused on the features and structures that afforded users *agency*, which is the ability to access and use resources to act on and influence existing environmental practices as well as create new practices (Norton, 2007). Enabling stakeholder agency is critical to development and implementation of successful environmental solutions (Norton, 2007). To identify the technological features and structures that enable and constrain user agency, I used DeSanctis and Poole's (1994) adaptive structuration theory (AST). Recognizing AST as a theory for businesses, I centered my analysis within the field of EPP by using Norton's (2007) environmental public participation framework.

Norton's framework uses stakeholder agency as a guideline for measuring the organizational mechanisms that lead to effective stakeholder participation. Although Norton's framework illuminates the organizational processes involved with designing and implementing EPP initiatives, it does not offer methodological considerations for identifying how technology, such as eco-apps, can increase stakeholder agency, which is why I coupled it with AST. After identifying and analyzing how the technological features of 11 eco-apps affected stakeholder agency, I offer a summary of the features and structures that EPP planners and scholars can use to create their own eco-apps. I conclude the chapter by discussing four best practices when designing eco-apps so as to increase stakeholder involvement and improve environmental outcomes.

Analytical Framework

I chose to use two theories grounded in Anthony Giddens's (1984) structuration theory because they provide means for analyzing the reciprocal influence of social structures and individual agency, a concept Giddens (1984) called *duality of structure*. Giddens (1984) explained the duality of structure as the process through which the rules and resources of social structures enable and constrain an individual's actions, while at the same time an individual's ability to reciprocally influence social structures comes from his or her agency, or ability to act freely within these structures. Norton (2007) adapted this concept for EPP, calling it *duality of participation* and explaining it as the process whereby stakeholder agency influences the reproduction or alteration of rules and resources, while at the same time rules and resources constrain stakeholder agency (Norton, 2007). For example, the structure of

EPP events concerning Colorado's public commentary on the Roadless Area Conservation Rule made it easy for revenue-focused groups to attend but harder for forest advocates to participate. Thus, the structure of the event privileged reproduction of discourse on economic gain over forest preservation (Smith & Norton, 2013). Web 2.0 technologies may offer a way to shift the duality of participation by offering stakeholders more agency in creation of the rules and resources of EPP initiatives. For example, stakeholders using Web 2.0 can create, coauthor, and control data through portals such as social networking sites (e.g., Facebook), video-sharing sites (e.g., YouTube), blogs (e.g., Pinterest), and wikis (e.g., Wikipedia) (Stern, 2015).

The co-creation process available via Web 2.0 is indicative of increased *discursive access*, which Norton (2007) describes as stakeholders' ability to exert agency by attaining and contributing to the environmental knowledge, materials, and procedures of an EPP initiative. Depoe and Delicath (2004) outline specific mechanisms for increasing stakeholder agency, including the ability to decide the mechanisms through which stakeholders participate, when to access information, and the ability to meaningfully contribute to the final product. The affordances of eco-apps using social media platforms offer the mechanisms Depoe and Delicath (2004) highlight by enabling stakeholders to contribute text, images, or video as well as control when and from where they contribute. Research shows that increased discursive access for stakeholders results in better decision-making efforts and improved scientific understanding of environmental problems (Fischer, 2000; Kasemir et al., 2003; Phillips et al., 2012). Alternatively, limiting stakeholder participation and feedback results in decreased understanding of environmental problems, fewer solutions, and less stakeholder buy-in (Phillips et al., 2012). The groups that organize EPP initiatives can restrict stakeholder agency by limiting the resources and discursive space available to stakeholders (Horsbøl & Lassen, 2012). In Peterson's (1997) account of public participation efforts aimed at squelching the spread of brucellosis disease in Canadian bison populations, she described how agency staff organizing the event privileged their technological discourse over the creative discourse of stakeholders through restricted meeting availability as well as by verbally discrediting and leaving out alternative perspectives in management decisions.

Yet simply increasing discursive access is not enough to ensure stakeholder participation in EPP initiatives. Norton (2007) raises Giddens's (1984) concepts of ontological security and competence as crucial to stakeholders' willingness and ability to meaningfully participate in initiatives. Giddens

(1984) argues that as people engage in daily routines, they gain *ontological security* or trust in existing social structures, but to act within those structures people must also have *ontological competence* or understanding of how they can act. Norton (2007) appropriates these concepts for EPP initiatives, explaining that ontological competence and security can increase participant agency in solving environmental problems. *Ontological competence* is crucial for both EPP stakeholders and organizers, as it ensures understanding of environmental problems and ways to structure EPP initiatives to collaboratively address those problems. *Ontological security* is established through processes and standards that build trust among EPP stakeholders and organizers that other parties' intentions are for the good of all. Some stakeholders in the Canadian bison brucellosis disease EPP initiative analyzed by Peterson (1997) could not gain ontological security because organizers placed their own interests at the forefront and limited stakeholder input. Additionally, the EPP organizers in Peterson's account lacked ontological competence in ways to accept and integrate the creative discourse of stakeholders as legitimate sources of knowledge on bison health.

Alternative ways of building ontological competence and security may come about through the features of coproduction and real-time feedback afforded via mobile phone applications using Web 2.0 technologies. For example, the process of working with others to solve environmental problems at times and in ways convenient for stakeholders may lead to increased ontological security and competence due to a more transparent structure and the contributions of others. Building trust and expertise across EPP planners and participations is important because it can lead to more effective knowledge production and public participation (Senecah, 2004; Kinsella, 2004). Eco-apps could also help build trust across participants through equal rhetorical presentation of conflicting stakeholder and agency perspectives (Schwarze, 2004). Additionally, by offering users space to identify "points of affiliation," they could further build trust by aiming for and reaching mutual goals (Schwarze, 2004).

The affordances of eco-apps may help overcome broader complications of in-person EPP initiatives, such as the permeability of environmental problems across political borders; costs associated with time, setup, and transportation; and a lack of systems to support adequate dialogue and incorporation of collaborative efforts (Depoe & Delicath, 2004; John, 2000; Savitz, 2000). The features of eco-apps offer a way for people in any location to upload data; the cost for development and use can be minimal; and newsfeeds can display the content of all contributors. Research has shown that the personalization, collaboration, and location features afforded by mobiles have influenced the movements, contributions, and knowledge of EPP stakeholders (Büscher, 2014; Newman et al., 2012; Shaw, Surry, & Green, 2015). The term *hybrid ecology* adequately addresses the uniquely mobile affordances of eco-apps, specifically their ability to augment users' understanding and experience of space, because they can retrieve previously inaccessible information (Licoppe, 2013). These examples point to the possibility that eco-apps improve EPP initiatives; however, it is not clear whether the features and structure of eco-apps support stakeholder agency.

To identify the eco-app features important to stakeholder involvement, I couple Norton's (2007) framework with DeSanctis and Poole's (1994) adaptive structuration theory (AST). AST originates from Giddens's (1984) structuration theory and Ollman's (1971) concept of appropriation and offers a methodological approach to understanding the way technology influences social structure and agency. Similar to Giddens's duality of structure and Norton's duality of participation, DeSanctis and Poole (1994) offer the *duality of technological structure,* which they describe as the intersection of the social structure built into technology and the social structure that emerges from users' appropriation of the technology. For example, a government agency may design a mobile app to encourage users to collect scientific data on tree populations; however, participants may use the app instead to discuss the decline of various tree populations.

DeSanctis and Poole (1994) gave specific concepts for analyzing each side of a technology's duality. The built technology offers specific social structures and action the user can potentially engage in, which is called the technology's *structural potential.* To analyze structural potential, a researcher identifies objectively describable *structural features* that can enable or constrain user agency, such as the rules, resources, and capabilities afforded by the technology. Next, the researcher surmises the goals, values, and action these features promote, which are called the *spirit* of the technology. However, once users interact with a technology, they may not engage in the social structures and action the designer intended, but rather *appropriate* the technology's features for their own social structures and action. Thus, researchers must also look for the ways users appropriate structural features, as well as the goals, values, and actions that come from users' appropriation.

The following analysis couples AST with Norton's framework by identifying the structural potential (RQ1) and user appropriation (RQ2) of

11 eco-apps as a means to increase stakeholder agency in their respective EPP initiatives. In the analysis, I focus specifically on Norton's components of user agency, specifically discursive access, ontological competence, and ontological security described previously.

Methods

I reviewed more than 100 eco-apps, both old and new, via newspaper articles, websites, and the iTunes and Google Play stores. I looked for applications that encouraged environmental public participation by prompting users to think, talk, or act for environmental change by co-creating content. Of all the apps reviewed, only 11 fit the above criteria (Table 12.1). These 11 apps fell into 2 structurally unique forms: 1) *native eco-apps*, designed specifically for environmental participation and requiring download from an app store, such as iTunes and Google Play, and 2) *embedded eco-apps*, which are eco-apps embedded in native apps not specifically designed for environmental participation.

Table 12.1. Eco-apps reviewed in this study

Native Eco-Apps	*Embedded Eco-Apps*
1. JouleBug	8. 3R Actions Challenge: @RecycleManiacs embedded in Twitter *(event ended)*
2. Marine Debris Tracker	9. BART Public Transport Badge embedded in Foursquare *(event ended and app discontinued)*
3. MyActions	10. Litterati: #litterati embedded in *Instagram*
4. Nature Near You *(discontinued)*	11. Sydney photo walk: #sydearthhourpw embedded in Instagram *(event ended)*
5. Project Noah	
6. Tracking the Wild	
7. WildLab Bird	

Procedure

In this project, I followed DeSanctis and Poole's (1994) recommendation for qualitative analysis, specifically to use Miles and Huberman's cross-case techniques (1994) to code each application for objectively describable features and the types of goals, values, and activities these features promoted. I developed my codes by creating a table where I listed the application names, features, and comments by users and developers. The table included information on the development of the eco-apps and structural potential of eco-app features, as well as data on how eco-app users appropriated those features.

I inferred the eco-apps' structural potential by looking at a range of resources, including news articles about the apps, the app creators' websites, content from app storefronts, and downloading and using the applications myself. I downloaded, explored, and used the apps on my Android phone and Apple iPad except for the BART badge, which was not accessible because of changes in the Foursquare application, and Tracking the Wild, 3R Actions Challenge, and #sydearthhourpw apps, which I could only explore because of my physical location or because the event had ended.

I documented user appropriation of eco-app structures by reviewing news articles about the applications, user comments on the app store, and user activity within the applications. According to DeSanctis and Poole (1994), users' postings within the applications offer a direct measure of their appropriation of the technology's features.

I engaged in an iterative process of reviewing the table and my sources as a way of understanding how the design of the apps offered users agency and the ways users appropriated technology to exert their agency. First, I identified the eco-app features that offered *discursive access* to stakeholders, such as those that offered a route for users to attain and contribute environmental knowledge, materials, and procedures. Then I noted how users appropriate those features to gather and post content. Second, I documented structural features that enabled the posting of multiple perspectives and collaborative opportunities—essentially any features that offered users a way to build *ontological competence* in the environmental topic and ways to solve it. Next I documented how users interacted with the aforementioned structural features to learn about environmental issues and take action to solve them. Third, I documented the eco-app processes and features that might help users build trust in one another and the software. To identify if users felt *ontological security* in the system, I recorded user comments posted in the apps and on the app store pages, specifically as to their enjoyment and appreciation of the app or lack thereof. Finally, I brought the findings together in such a way so as to explain the features in eco-apps that afforded user agency (research question one) and how users engaged with these features to employ their agency (research question two).

Analysis

Comparing the structural potential to the appropriation of eco-apps, I found congruency across the goals of app creators and the actual behaviors of end users. The eco-apps offered users agency through the structural features of visualizations, community forums, games, data, and interactions with their environment. Users commented on ways these structures increased their agency in and ontological competence of environmental problems and what they could do to solve them. Ontological security varied: some users appreciated the app and even knew the creators, while other users found that numerous bugs and constant crashes made their attempts to participate a challenge.

Research Question 1

In answering research question 1, *What is the structural potential of specific eco-apps to increase stakeholder agency?*, I found the following features assisted in offering the structural potential to increase stakeholder agency: tagging, posting, commenting, liking, sharing, badges, points, missions, and location-based prompts. These features offered the structural potential for users to exert agency by gathering, creating, and sharing environmental knowledge through the structural features of visualizations, community forums, games, data, and interactions with their environment.

All of the apps reviewed used *visualizations* to coalesce user-generated content on dashboards, newsfeeds, tables, and maps. Examples included a table in MyActions showing the total amount of water conserved by users and a world map in #Litterati showing locations of user-collected litter. Creators of MyActions describe on their Google Play store page the importance of visualizations to users of their app, "Measure Your Impact—Visualize the real time impact of personal and campus actions." Tracking the Wild promoted ontological competence of African wildlife by offering a link to descriptions of animals on Wikipedia in the same tab that allowed for uploading and identification of wildlife sightings. Users in Tracking the

Wild and The WildLab Bird apps could also learn about wildlife or birds
from other users' posts and posts' geotags (Figure 12.1). Apps reflected user
competence and discursive access through personalized visualizations of their
contributed content.

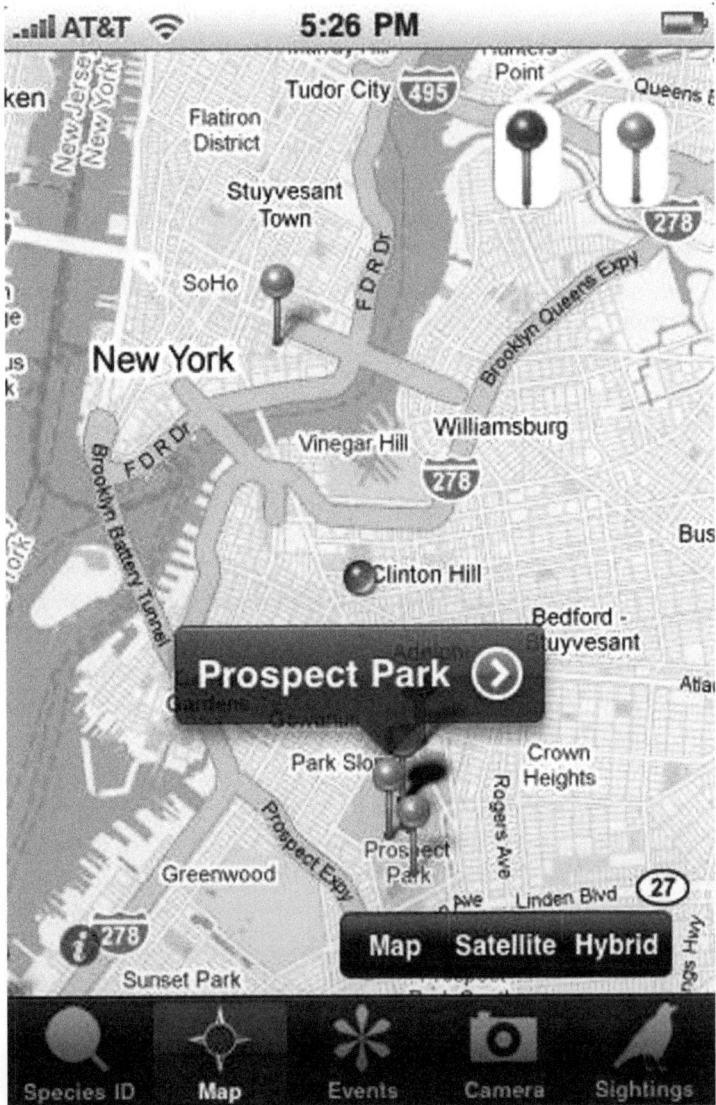

Figure 12.1. The WildLab Bird map, which enables users to see each other's geotagged posts.

While in the Tracking the Wild app, users could see the parks they visited and animals recorded. In the JouleBug app, users could see the total amount of money and carbon dioxide they saved from participating in sustainability challenges (Figure 12.2). Visualizations appeared to offer both

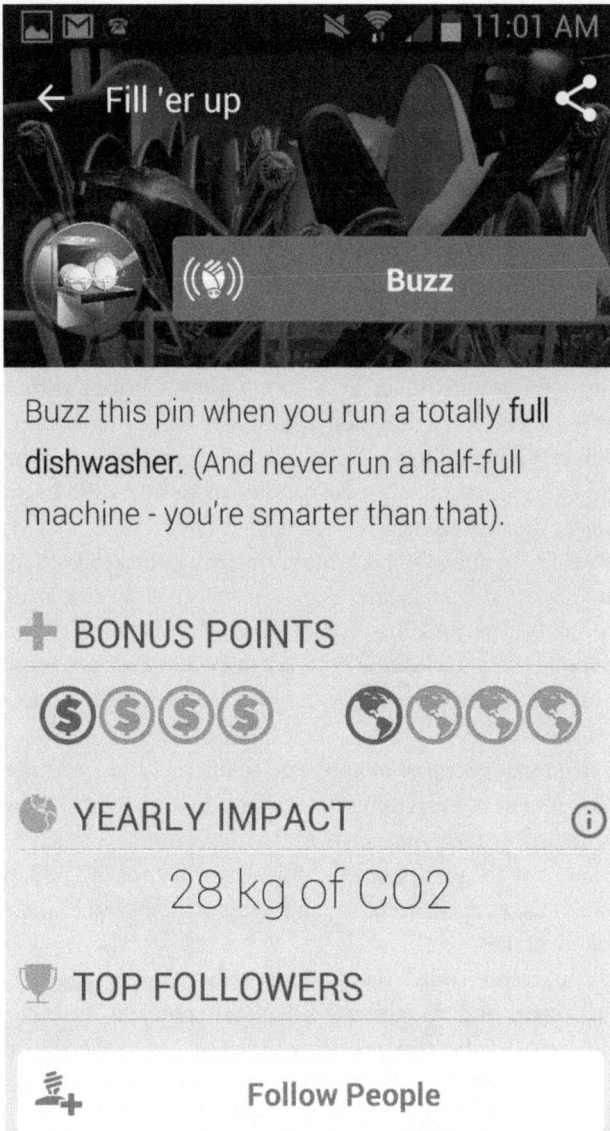

Figure 12.2. Visualization of the CO_2 saved by participating in a JouleBug challenge.

ontological competence, as described, and ontological security by showing the efforts of other users and highlighting the importance of each contribution. Creators of Tracking the Wild described on their iTunes page the value of user postings to conservation efforts, even citing a University of Cape Town professor's use of the data to offer proof of the need to protect specific species.

Community forums included commenting, liking, and sharing features where participants posted ideas, made suggestions, shared content, and organized offline events. All of the apps included some aspect of community forums, but Project Noah, WildLab Bird, MyActions, and JouleBug encouraged users to meet in person by offering ways they could create, share, or participate in off-line events or missions in their community.

Data, such as Wikipedia pages, bird sounds, nature news, and conservation tips, offered descriptions and information to build app users' ontological competence. Five of the 11 eco-apps, including JouleBug, MyActions, Nature Near You, Tracking the Wild, and WildLab Bird, offered environmental data. Often apps coupled data with *gamification* features, such as badges, points, actions, check-ins, missions, contests, and competitions. Five of the 11 eco-apps, including JouleBug, MyActions, Project Noah, BART Badge, and @RecycleManiacs, used features to gamify EPP. In an effort to reduce carbon dioxide emissions related to transportation, Bay Area Rapid Transit (BART) "gamified" check-ins by offering three $25 transit tickets to randomly selected Foursquare users who checked in at a BART station in January 2010. The JouleBug app (Figure 12.2) coupled game elements, such as competitions against one's social network for badges and points, with data about the impact of the user's actions toward reducing carbon dioxide emissions.

The structural potential of apps offered another route for users to build ontological competence regarding nature and their own actions by prompting *interactions with their environment.* These prompts included suggestions for engagement with users' homes, offices, and nature. The #Litterati app (Figure 12.3) encouraged users to find, photograph, and dispose of litter they found in nature.

The app accompanied the prompt with a *visualization* of the litter collected by others and the ability to like and comment on each post. The @RecycleManiacs, embedded in Twitter, encouraged university students to reduce, reuse, and recycle at home and at their universities through competitions with other universities. On their website, creators of WildLab Bird described the importance of users' acting as stewards of their environment

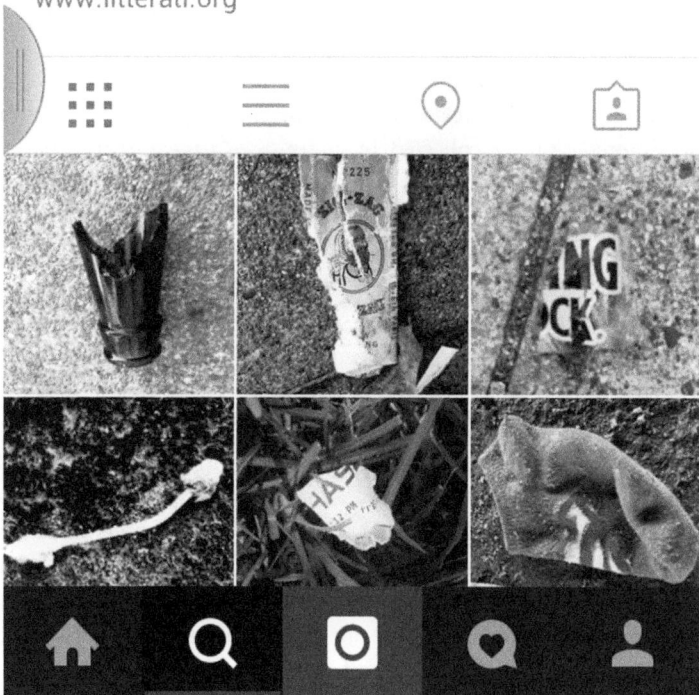

Figure 12.3. Visualization of litter collected.

by identifying and posting their bird sightings, while creators of MyActions explained on their Google Play story the importance of users following app prompts to "experience" their environment and "to inspire others to do more."

Research Question 2

In answering research question 2, *How did users of these eco-apps appropriate the technology to act on and influence existing environmental practices as well as create new practices?*, I found users appropriated the features of eco-apps in accordance with the structural potential as described under research question 1. Users exerted agency by accessing and using apps to engage in the suggested environmental practices; however, few users used app features to create new practices or attend in-person events. Additionally, the differing structures of embedded and native apps affected how often and how much users appropriated the technology. Users' ontological competence and security appeared relatively high based on their comments; however, challenges with the app interface crashing and low user numbers appeared to damage users' access to knowledge and trust in eco-apps.

The *visualizations* of users' contributions offered a way to document appropriation and discursive access across users, with examples such as nearly 400,000 pieces of litter collected by users of Marine Debris Tracker and 32,000 items diverted from landfills by RecycleMania users. Additionally, the BART Foursquare app recorded increased ridership by 19% and frequency of rides by 14%, and 38% of riders felt the app made BART more fun to ride (Mager, 2010). Some of the apps did not offer visualizations showing the exact quantity of contributions; instead, they showed user contributions through pins on a map, image newsfeeds, or comments, as well as individual user profiles.

To better understand appropriation, I looked at posts and noted the frequency and quantity of activity across apps, which revealed a difference between appropriation of embedded and native apps. The users of native apps appeared to post more content more often, while users of embedded apps appeared to post less content less often. Upon examination of users' profiles, it appeared that native app users appropriated the applications for a longer period of time or duration than embedded apps users. For example, the native app users of JouleBug and MyActions reported their daily and weekly activity, while users of Marine Debris Tracker, Nature Near You, Project Noah, and WildLab Bird posted 10 to 50 times within a one- to two-hour time frame about once a month. The users of embedded apps posted about one to five times total, with the exception of #Litterati, where users posted one to five times per month.

The difference in appropriation between the apps was most evident when I compared a native and embedded app requesting the same type of participation: litter collection. The native app Marine Debris Tracker requested that users record the type and quantity of litter, then throw it away. The embedded app #Litterati requested that users photograph litter, then throw it away. Marine Debris Tracker had a dedicated user base, recording litter collection of 10 to 50 articles weekly, while #Litterati users posted one to five times per month. Reflecting on how the structure of the apps may have affected user activity, the embedded app, on one hand, offered one simple feature: take a picture within a native app, Instagram, that many people already used for other purposes.

On the other hand, Marine Debris Tracker required that users search for the app in an app store, download the app, fill in detailed information (Figure 12.4 on page 318), and remember to use it, because it is not an app they use for other purposes.

To get a sense of users' *ontological competence* and *security*, I recorded user comments on app store pages and in the apps. In line with app creators' visions (see research question 1), users wrote that they learned from and appreciated the features and design of eco-apps. On the Project Noah iTunes page, app user Trailerville explained the importance of the app's locative features, community forums, gamification, and prompts for interaction with nature in his quest to document wildlife. Trailerville wrote, "My in-field photo documentations are automatically GPS and date stamped and I was able to create a 'mission' for the group that has shared our reptile findings for the past 11 years. Now suddenly our work = a database, a shareable body of knowledge, available to the world." On the Marine Debris Tracker GooglePlay store page, an anonymous user described her appreciation for the app and her relationship with the app creator: "Love it! Big advocate of the app, its functionality and its creator! Jenna is easy to work with and this app has the potential to be used for cleanup groups to compile their data! Saves so much time and money! Spreading it on the West Coast! Contact her directly to help with edits and shes on it or she can tell you if it cannot be done ;)"

Commenting specifically on gaining ontological competence and showing ability to exert agency in the EPP initiative, Hannah Maxwell wrote on the JouleBug GooglePlay Store app page, "Great way to learn. I love using

View and Submit
2 Items

Lat,Lng =
(37.3307991027832,-122.03073120117188)

Accuracy < 10m

Top Items ▲

Aluminum or tin cans	-	1	+	Log
Plastic Bags	-	1	+	Log
Plastic Bottle	-	1	+	Log
Glass Bottle	-	1	+	Log
Plastic Bottle or Container Caps	-	1	+	Log
Cigarettes	-	1	+	Log
Plastic Food Wrappers	-	1	+	Log
Plastic Utensils	-	1	+	Log

Figure 12.4. Upload screen for Marine Debris Tracker.

this app to teach sustainability to my classes! It helps keep my bills low and the air clean." Project Noah user Irishpan explained on the app's iTunes page the importance of the app in building his son's ontological competence and discursive access. "For the last year, I've been doing a leaf collection with my son to help him learn to notice, describe, and recognize things around him. But it really only worked for plants . . . Then we discovered Noah. Not only does it teach him the above, it has revitalized his interest. He loves knowing that he is helping real scientists and if Mommy can't find out what it is, someone else can help us figure it out. Earning patches offers a great set of goals. And best of all, we can surf other people's amazing sightings. Incredible project! Wonderful app!" WildLab Bird also noted on its webpage the contribution of children and appreciation of teachers toward its conservation efforts. "Students enjoyed the WildLab program, found the iPhone app helpful, easy, and fun, demonstrated increased content knowledge, and had increased interest in studying science and pursuing careers in science after the 5-session program. Every teacher that participated said they would participate again if offered the opportunity."

One structural design users did not appropriate as anticipated was the community forums to develop new practices or meet in person. Eco-app users typically appropriated the commenting, liking, and sharing features of the apps to show their actions to the community and offer community support; however, they did not use these features often to post ideas, make suggestions, or organize off-line events. Some exceptions include several users on MyActions who offered their community creative sustainability practices and events to attend. JouleBug offered badges people could complete in their community, but did not specifically offer these as opportunities for discussing ways to address environmental problems. Finally, Project Noah and WildLab Bird garnered activity for in-person events via teachers who organized students for outings to document wildlife. These examples point to an inadequacy in the current eco-app design and its appropriation by users, specifically in the area of stimulating new practices and collaborative discussions to solve environmental problems.

Complications with the interface further hampered users' *ontological security* and ability to engage in *discursive access*. On the MyActions Google Play store page, Benjamin Leamon commented on his difficulty in getting the app to work. "Slow, Clunky, Buggy. This app is barely more than a link to the myactions web site. I cannot upload photos (even after the update!), the site is prone to crashing, and many of the links just don't work on my device." On the Tracking the Wild Google Play store page, Sarien Lategan wrote about her frustrations with not being able to use the app. "Not working. I was extremely excited about this app but can't get it to work properly. It keep force shutdown and not all the functionalities seem to be there."

The above analysis shows the great potential for eco-apps to serve as a digital space for increasing discursive access, ontological security, and competence of environmental public participation stakeholders. Users of the eco-apps readily appropriated the apps' structural features, yet some of the more crucial features that could lead to deeper understanding and problem solving of environmental problems were not well appropriated.

Conclusion: Best Practices for Future Efforts

It appeared that the construction of eco-apps and user appropriation followed DeSanctis and Poole's (1994) ideal *duality of technological structure* because the social structure built into eco-apps closely mirrored the social structure that emerged from users' appropriation of the apps. When app structure offered users a way to gamify sustainable activities or visualize the amount of litter collected, users participated and rarely deviated to use the app for other purposes. Thus, from the perspective of DeSanctis and Poole, the small deviation in the duality of structure lead to countless conservation efforts in accordance with eco-app creator's intentions.

Framing the eco-app analysis with Norton's (2007) *duality of participation* reveals relatively strong user agency. Users of the eco-apps acquired information and contributed content through a range of features, including tagging, posting, commenting, liking, sharing, badges, points, missions, and location-based prompts. These features offered various structures that users could engage in EPP, including visualizations, community forums, games, data, and interactions with their environment. Users affected the overall structure of the app through the types of content and how often they uploaded content. Conversely, defects in the app structure, specifically bugs, crashes, and lack of functionality across different mobile devices, constrained stakeholder agency. Notwithstanding the defects, users participated in numerous environmental activities through the apps, such as sustainable behaviors and citizen science.

However, the types of participation the app developers created and users engaged in afforded insight to only one dimension of the multidimensional environmental problem they hoped to solve. This *one-dimensional environmental public participation* limited user agency and ontological competence to one variable of the environmental problem instead of offering avenues for users to explore and realize other variables causing the problem. Alternatively, if eco-apps offered a *multidimensional environmental public participation* experience, stakeholder participation could uncover other variables, such as social, political, and economic dimensions affecting the environmental problem and solutions to that problem (Figure 12.5).

For example, the JouleBug app instructed users to "kill the lights when you leave a room" to conserve 200 kg of carbon dioxide per year. Yet this one-dimensional "solution" to carbon dioxide leaves out the myriad factors contributing to the environmental problem, such as leaving lights on to deter home intruders, government subsidies to polluting power sources, and the relative cheapness of electricity. Similarly, Tracking the Wild instructed African wildlife tourists to photograph and identify animals to assist conservation efforts, while neglecting the social, political, and economic factors creating the need to protect African wildlife from poachers and habitat destruction. These examples show that even though eco-apps engaged stakeholders, they

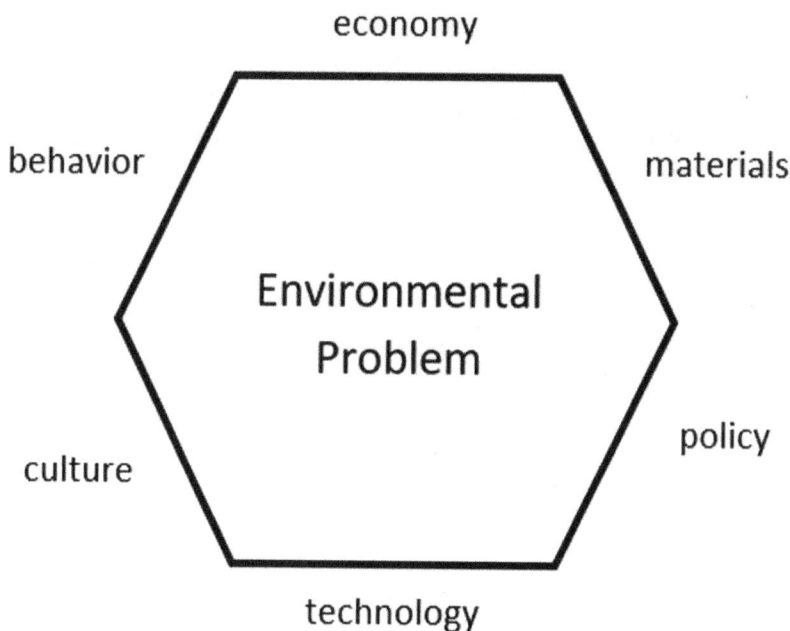

Figure 12.5. Multiple dimensions to environmental problems.

still lacked the ability to bring together more diverse knowledge, perspectives, and values needed to enact insightful public conversation and environmental change (Kinsella, 2004).

The structural potential for users to engage in multidimensional EPP exists in eco-apps, yet this potential was not appropriated or realized by users. For example, MyActions offered users the ability to like, comment, thank, and share unique content while on the go as well as meet in person to start dialogues that could lead to multidimensional understanding. Yet the users of MyActions stuck to the suggested actions of taking public transit, recycling, and conserving water. They did not use the features of the app to offer ways to address why public transit doesn't reach everyone, find solutions to the pollution caused by recycling, or question the reason behind water shortages. Büscher and Igoe (2013) argue that the features of Web 2.0 lead users to believe the simple acts of liking and sharing are effective forms of environmental participation, thus leading to an artificial understanding of relationships causing environmental problems. Contrary to Buscher and Igoe (2013), I do not think the features of Web 2.0, such as liking and sharing, are to blame for a lack of understanding or that they inherently lead to one-dimensional EPP. Rather, the execution of these features in the design of technological interfaces segues users into specific paths of appropriation and, thus, either limits or broadens user understanding. In the case of MyActions, the suggested actions appeared most evident in the interface, while the ability to use the features for deeper exploration was less obvious to users.

My contribution to the discussions of EPP and technology is to highlight the need to reconceptualize Web 2.0 technologies, such as eco-apps, as pathways to multidimensional EPP. If EPP organizers and technology developers start with the intention to capture, connect, and explore the multiple variables of environmental problems (see Figure 12.5), the structural potential and appropriation of technology may look radically different. I offer the following four best practices to move the development of eco-apps toward a more *multidimensional environmental public participation* structure:

1. Include and support diverse stakeholders in the *development* of eco-apps. The execution of features into specific structures, and eventually appropriation by users, relies directly on the people developing the technology. The more diverse the stakeholders participating in the development of the technology, the higher the chance that the technology will include more than one dimension of an environmental problem. This could mean opening brainstorming sessions for the design of eco-apps to stakeholders or assisting stakeholders in creating their own apps that work together to show different dimensions of the issue while at the same time moving users toward truly addressing the environmental problem as a whole.

Although I did not cover developer inputs as part of my formal analysis in this paper, I dedicated several columns in my spreadsheet to tracking the types of developers and development processes. I reserved these columns because of curiosity stemming from Norton's (2007) argument that for an EPP event to be successful, organizers must also hold some ontological competence in organizing these events. Drawing information from developers' websites, app store pages, and news articles, I found that the resources required to develop native and embedded eco-apps impacted the structural potential of the app. On the one hand, native apps were developed in partnership with nonprofits, government agencies, and/or the private sector and required people proficient in coding and design. The structural potential of the resulting native apps focused on continuous, long-term use by relatively dedicated users. For example, the native app JouleBug, created by a team of entrepreneurs and programmers, aimed to help users live a sustainable lifestyle by providing badges and pins that coincided with the time of year or current events. On the other hand, embedded apps were developed by one organization or one person and required no knowledge of coding or design, only that the developer know how to create and publicize a hashtag within the native app and his or her off-line community. The structural potential of the resulting embedded apps focused on spontaneous, short-term use by casual users. For example, the #sydearthhourpw app, created by the World Wildlife Fund, ran for one day and asked users to snap a photo of a street scene exemplifying what a sustainable Sydney might look like.

The two types of apps, native and embedded, offered a range of possibilities for integration of stakeholders in the development processes because they both use different levels of ontological competence related to app construction. The first step of eco-app design is to gather diverse stakeholders to discuss the many dimensions of an environmental problem. Upon identifying these dimensions, the second step entails stakeholders deciding how technology might assist in addressing the environmental problem. One possibility is for stakeholders to contribute time, money, and knowledge to the development of a native app that reveals the many dimensions of the problem. Another possibility is for stakeholders to agree to create and maintain their own embedded app, which would reflect one or two of the dimensions identified, as well as connect to the other embedded apps so as to bring users to a fuller understanding of the environmental problem. Either way, the process of bringing stakeholders together to discuss the social, economic, and political causes of the problem will serve as an EPP initiative within itself, further solidifying ontological security in the participants and widening stakeholder's ontological competence in the topic. Thus, incorporating multiple stakeholders in the planning of an eco-app from its inception can lead to multidimensional understanding of the problem and a technological structure that accommodates that understanding.

2. Incorporate diverse stakeholders as *users* of eco-apps. Because Web 2.0 features offer a way for users to co-create content, the more diverse stakeholders using the technology, the more likely users will identify and address multiple dimensions of the environmental problem. Following the first suggestion, the inclusion of stakeholders in the development of an eco-app provides an opportunity to identify possible roles and contributions of a range of stakeholders. With a range of contributors, an eco-app developer could then develop space for those contributions into the structure of the app, which could lead to better appropriation by users. Therefore, the design of the app offers ways for multiple stakeholders to become involved, such as those from government agencies, nonprofits, universities, business, and so forth. By involving multiple stakeholders in eco-apps, users can present a diverse array of local knowledge and expertise, which can address current challenges to increasing publicly available knowledge (Kinsella, 2004).

Drawing from my analysis, several apps aimed to incorporate other stakeholders as an outcome of the app's use. #Litterati's creator, Jeff Kirschner, said, "The idea [of #Litterati] is to build this grassroots campaign, then using that who, what, where, and when data, go top-down and work with cities to be more strategic about the placement of trash cans and recycling bins; or work *with* brands to be more strategic about the type of packaging that they create in *hopes of* eventually reaching a litter free planet" [emphasis in original] (Kirschner, 2013). This quote shows that Kirschner does not assume the only purpose of his embedded eco-app is to collect and document litter; rather, he describes user data as a way to start a dialogue with the organizations that have power to change production practices and the agencies that can influence citizen behavior. Similarly, the World Wildlife Fund, creators of #sydearthhourpw embedded eco-app, asked users to upload images "to re-imagine the modern sustainable city," and, with such success from the first, they created a second embedded eco-app, #abettersydney, that asked users to photograph the possibilities for a car-free street through

Sydney. These images were then taken to a Sydney city council meeting to challenge council members to build a sustainable Sydney (Myers, 2011).

The creators of both #Litterati and #sydearthhourpw aimed to integrate other stakeholders in solving environmental problems, but neither invited these stakeholders to literally participate in and become users of their embedded apps. The outcomes of the apps may have been different had the creator of #Litterati invited businesses to participate, especially those whose products ended up as litter. Similarly, if the developers of #sydearthhourpw invited local businesses to participate, they may have found more dimensions of the problem, such as the capacity and interest of local businesses to contribute to the effort. The successful integration of other stakeholders in the interface relies heavily on their involvement in the development stage, which offers a chance for all parties to build ontological security among one another and competence in ways to solve the problem. Simply offering a place for stakeholders after the development of the app will not ensure their participation in the appropriation of the app.

3. Record lived experience and live new experiences through eco-apps. Designing eco-apps to enable the recording of experiences and to prompt new experience in nature can open the door to multidimensional understanding of environmental problems. Nothing is static in human-nature interactions; mobiles offer an easy way to uncover, document, and share users' constantly changing interactions with their environment. Users can develop ontological competence of place by reporting what they observe and going on missions to discover that which they have not yet seen or experienced.

The eco-apps in this study offered ways for users to record their lived experience. Nature Near You eco-app users took photos and marked noteworthy locations and wildlife in parks, while users of the #sydearthhourpw were asked to explore Sydney and post images of what a sustainable city might look like. In the process of exploration, users can develop ontological security through reflecting on their contributions in relation to the contributions of other users. For example, the WildLab Bird app allowed users to reflect on their location in relation to bird sightings and neighborhood events posted by other users while also projecting their bird sightings on a map.

Connecting location and temporality (such as seasonal changes) with the ability to co-create content allows users to pin new meaning to places through digitally labeling or tagging places with descriptions and images of wildlife or geographic features. The discursive access offered through uploading one's own experience can lead to the development of place meanings, which are a person's beliefs, feelings, memories, and knowledge about a place (Cheng et al., 2003). By helping users develop place meanings within their community, mobiles can influence stakeholders' attitudes towards changes, both positive and negative, at that place (Jacobs & Buijs, 2011). When structuring eco-apps, developers should consider ways to develop the five types of place meanings through user experience: beauty (aesthetic judgments), functionality (ways of use), attachment (feelings of belonging), biodiversity (meanings pertaining to nature), and risk (worries about current or future events) (Jacobs & Buijs, 2011). As users explore and develop place meanings, they have the opportunity to learn for themselves how social, political, and economic structures influence conservation and sustainability practices.

4. Supplement and support offline relationships through eco-apps. By supplementing and supporting stakeholders' offline relationships, eco-apps could help stakeholders recognize the human dimensions assisting and hindering solutions to environmental problems. This might mean using eco-apps as a tool to coordinate in-person meetings, offering resources to supplement in-person meetings, and supporting activity that addresses emergent aspects of environmental problems. As a support tool, eco-apps could assist in building ontological competency in the ways communities can address environmental problems, as well as increasing users' ontological security in other stakeholders.

Some of the eco-apps in this study attempted to physically bring users together, such as WildLab Bird's events tab and MyActions' request for users to post community events; however, not many users appropriated these features. However, if these app developers placed more emphasis on in-person events within the interface, the apps may have functioned like this:

- Events on WildLab Bird connect users with biologists at a community outing to discuss what local activities threaten the survival of birds, which could generate new experiences for users and new relationships with community members.

- #Litterati could incorporate monthly litter cleanups at specific locations where users meet city staff and discuss the challenges of acquiring and locating trash bins for reducing litter.

In addition to holding events, eco-app developers could connect users in person through *gamification*. Gamification is the process of applying game design elements, such as points, badges, avatars, and leaderboards, to ordinary tasks so as to make these tasks more engaging and gamelike (Liu & Santhanam, 2015). Gamification offers an alternative way for users to meet other stakeholders through competitions for badges, missions, and points. From this study, JouleBug gamified opportunities to meet other stakeholders by offering users missions within the user community, such as volunteering in the community. Other ideas for gamification include:

- mayorship titles for those who perform the most environmental actions at one location

- increased points for people who attend in-person events

- a swarm badge (like that used on the original Foursquare) to award points only to users after a set number of diverse stakeholders arrives at a specific location

By instigating off-line encounters, eco-apps could increase the probability of users to observe and reveal the social, economic, and political practices causing environmental problems.

The content of this chapter provides important implications for technology use, specifically mobile phone applications, in the research and enactment of environmental public participation (EPP). This analysis revealed that to get at the root of environmental problems, we need to reconsider how technologies can facilitate exploration of the multiple dimensions of those problems. The most important step in this process is to include diverse stakeholders, such as businesses, nonprofits, and concerned citizens, before any technology is designed or developed. Starting technology development with public participation will, in turn, increase the likelihood that (1) more of the public will participate in using the app, (2) more dimensions of the problem will be revealed, and (3) mobile phones serve a key role in supporting solutions to complex environmental problems.

References

Büscher, B. (2016). Nature 2.0: Exploring and theorizing the links between new media and nature conservation. *New Media & Society, 18*(5), 726–743. doi:10.1177/1461444814545841

Büscher, B., & Igoe, J. (2013). "Prosuming" conservation? Web 2.0, nature and the intensification of value-producing labour in late capitalism. *Journal of Consumer Culture, 13*(3), 283–305. doi:10.1177/1469540513482691

Castells, M. (2007). Communication, power and counter-power in the network society. *International Journal of Communication, 1*(1): 238–266.

Cheng, A. S., Kruger, L. E., & Daniels, S. E. (2003). "Place" as an integrating concept in natural resource politics: Propositions for a social science research agenda. *Society of Natural Resources, 16*(2), 87–104.

Depoe, S. P. (2004). Public involvement, civic discovery, and the formation of environmental policy: A comparative analysis of the Fernald citizens task force and the Fernald health effects subcommittee. In S. P. Depoe, J. W. Delicath, & M. A. Elsenbeer (Eds.), *Communication and public participation in environmental decision making* (pp. 157–73). Albany, NY: State University of New York Press.

Depoe, S. P., & Delicath. J. W. (2004). Introduction. In S. P. Depoe, J. W. Delicath, & M. A. Elsenbeer (Eds.), *Communication and public participation in environmental decision making* (pp. 1–11). Albany, NY: State University of New York Press.

DeSanctis, G., & Poole, M. S. (1994). Capturing the complexity in advanced technology use: Adaptive structuration theory. *Organization science, 5*(2), 121–147.

Fischer, F. (2000). Preface. *Citizens, experts, and the environment: The politics of local knowledge* (pp. ix–xiv). Durham, NC: Duke University Press.

Fredericks, J., & Foth, M. (2013). Augmenting public participation: Enhancing planning outcomes through the use of social media and web 2.0. *Australian Planner, 50*(3), 244–256.

Giddens, A. (1984). *The constitution of society: Outline of the theory of structuration.* Cambridge, MA: Polity Press.

Horsbøl, A., & Lassen, I. (2012). Chapter 8: Public engagement as a field of tension between bottom–up and top–down strategies: Critical discourse moments in an "Energy Town." In L. Phillips, A. Carvalho, & J. Doyle (Eds.), *Citizen voices: performing public participation in science and environment communication* (pp. 163–186). Chicago, IL: Intellect.

Jacobs, M. H., & Buijs, A. E. (2011). Understanding stakeholders' attitudes toward water management interventions: Role of place meanings. *Water Resources Research, 47*(1). doi:10.1029/2009wr008366

Jentoft, S., & Chuenpagdee, R. (2009). Fisheries and coastal governance as a wicked problem. *Marine Policy, 33*(4), 553–560.

John, D. (2000). Good cops, bad cops. In J. Cohen & J. Rogers (Eds.), *Beyond backyard environmentalism* (pp. 61–64). Boston, MA: Beacon Press.

Kasemir, B., Jager, J., & Jaeger, C. C. (2003). Chapter one: Citizen participation in sustainability assessments. In B. Kasemir, J. Jager, C. C. Jaeger, & M. T. Gardner (Eds.), *Public participation in sustainability science: A handbook* (pp. 3–36). Cambridge, UK: Cambridge University Press.

Kinsella, W. (2004). Public expertise: A foundation for citizen participation in energy and environmental decisions. In S. P. Depoe, J. W. Delicath, & M. F. A. Elsenbeer (Eds.), *Communication and public participation in environmental decision making* (pp. 83–98). Albany, NY: State University of New York Press.

Kirschner, J. (2013, Mar 27). *Litterati—Where it all started* [Video file]. Retrieved from https://www.youtube.com/watch?v=4GxGJ8UefiI

Licoppe, C. (2013). Merging mobile communication studies and urban research: Mobile locative media, "onscreen encounters" and the reshaping of the interaction order in public places. *Mobile Media & Communication, 1*(1), 122–128. doi:10.1177/2050157912464488

Liu, D., & Santhanam, R. (2015). Towards meaningful engagement: A framework for design and research of gamified information systems. *Social Science Research Network*. Retrieved from http://papers.ssrn.com/sol3/Papers.cfm?abstract_id=2521283

Mager, A. (5 May 2010). How mass transit integrates with Foursquare. *ZDNet*. Retrieved from http://www.zdnet.com/article/how-mass-transit-integrates-with-foursquare/

Miles, M. B., & Huberman, M. (1994). *Qualitative data analysis: An expanded sourcebook*. London: Sage.

Myers, C. B. (2011, August 30). Earth hour global uses Instagram to promote environmental consciousness. *TNW News*. Retrieved from http://thenextweb.com/socialmedia/2011/08/30/earth-hour-global-uses-instagram-to-promote-environmental-consciousness/#

Newman, G., Wiggins, A., Crall, A., Graham, E., Newman, S., & Crowston, K. (2012). The future of citizen science: emerging technologies and shifting paradigms. *Frontiers in Ecology and the Environment, 10*(6), 298–304. doi:10.1890/110294

Norton, T. (2007). The structuration of public participation: Organizing environmental control. *Environmental Communication, 1*(2), 146–170.

Ollman, B. (1971). *Alienation: Marx's conception of Man in capitalist society*. Cambridge, UK: Cambridge University Press.

O'Reilly, T. (2005). What is Web 2.0: Design patterns and business models for the next generation of software. Retrieved from http://oreilly.com/web2/archive/what-is-web-20.html

Peterson, T. R. (1997). Subverting the culture of expertise: Community participation in development decisions. In T. R. Peterson, *Sharing the Earth: The rhetoric of sustainable development* (pp. 86–118). Columbia, SC: University of South Carolina Press.

Phillips, L., Carvalho, A., & Doyle, J. (2012). Chapter 1: Introduction. In L. Phillips, A. Carvalho, & J. Doyle (Eds.), *Citizen voices: Performing public participation in science and environment communication* (pp. 1–18). Chicago, IL: Intellect.

Reynolds, G. (2006). *An army of Davids: How markets and technology empower ordinary people to beat big media, big government and other goliaths*. Nashville, TN: Nelson Current.

Savitz, J. (2000). Compensating citizens. In J. Cohen & J. Rogers (Eds.), *Beyond backyard environmentalism* (pp. 65–69). Boston, MA: Beacon Press.

Schwartz, R. (2014). Online place attachment: Exploring technological ties to physical places. In A. de Souza e Silva & M. Sheller (Eds.), *Mobility and locative media: Mobile communication in hybrid spaces* (pp. 85–100). New York, NY: Routledge.

Schwarze, S. (2004). Public participation and (failed) legitimation: The case of forest service rhetorics in the boundary waters canoe area. In S. P. Depoe, J. W. Delicath, & M. A. Elsenbeer (Eds.), *Communication and public participation in environmental decision making* (pp. 137–156). Albany, NY: State University of New York Press.

Senecah, S. (2004). The trinity of voice: The role of practical theory in planning and evaluating the effectiveness of environmental participatory processes. In S. P. Depoe, J. W. Delicath, & M. A. Elsenbeer (Eds.), *Communication and public participation in environmental decision making* (pp. 13–34). Albany, NY: State University of New York Press.

Shaw, E. L., Surry, D., & Green, A. (2015). The use of social media and citizen science to identify, track, and report birds. *Procedia—Social and Behavioral Sciences, 167*, 103–108. doi:10.1016/j.sbspro.2014.12.650

Shirky, C. (2008). *Here comes everybody: The power of organizing without organizations.* New York, NY: Penguin Books.

Smith, H. M., & Norton, T. (2013). "That's why I call it a task farce": Organizations and participation in the Colorado Roadless Rule. *Environmental Communication: A Journal of Nature and Culture, 7*(4), 456–474.

Stern, J. (2015). Introduction to Web 2.0. *West Los Angeles College.* Retrieved from http://www.wlac.edu/online/documents/Web_2.0%20v.02.pdf

Tapscott, D., & Williams, A. (2006). *Wikinomics: How mass collaboration changes everything.* New York, NY: Penguin Books.

Walker, G. B., Daniels, S. E., Blatner, K. A., & Carroll, M. S. (1996). *Civic discovery and ecosystem-based management: Collaborative learning in fire recovery planning.* Paper presented at the Speech Communication Association Convention, San Diego, CA.

List of Contributors

Steve Daniels is Professor, Department of Sociology, Utah State University.

Stephen P. Depoe is Professor and Head, Department of Communication, University of Cincinnati.

Giles Dodson is Senior Lecturer, School of English and Media Studies, Massey University (NZ).

Danielle Endres is Professor, Department of Communication, University of Utah.

Susan Hansen is a former Delta County (CO) administrator (retired) and cattle rancher.

David D. Hart is Director, George J. Mitchell Center for Sustainability Solutions, University of Maine.

Jill E. Hopke is Assistant Professor, College of Communication, DePaul University.

Kathleen P. Hunt is Assistant Professor, Department of Communication, State University of New York-New Paltz.

Colene J. Lind is Assistant Professor, Department of Communication Studies, Kansas State University.

Carmine Lockwood is a Renewable Resources Staff Officer (retired), Grand Mesa, Uncompahgre, and Gunnison National Forests (GMUG) of Colorado.

Bridie McGreavy is Assistant Professor, Department of Communication and Journalism, University of Maine.

Matthew McKinney is Director, Center for Natural Resources and Environmental Policy, University of Montana.

Nicholas P. Paliewicz is Assistant Professor, Department of Communication, University of Louisville.

Anna Paliser is Polytechnic Tutor, Environmental Management, Southern Institute of Technology (NZ).

Tyler Quiring is a doctoral student, Department of Communication and Journalism, University of Maine.

Lydia Reinig is a doctoral student, College of Media, Communication and Information, University of Colorado.

Susan Senecah is Professor Emerita, State University of New York College of Environmental Science and Forestry and long-time consultant with the New York State Legislature.

Linda Silka is Senior Fellow, George J. Mitchell Center for Sustainability Solutions, University of Maine.

Molly Simis-Wilkinson, Ph.D., is an independent researcher.

Stacey K. Sowards is Professor and Chair, Department of Communication, University of Texas at El Paso.

Leah Sprain is Assistant Professor, College of Media, Communication and Information, University of Colorado.

Chui-Ling Tam is Assistant Professor, Department of Geography, University of Calgary (CA).

Carlos A. Tarin is Assistant Professor, Department of Communication, University of Texas at El Paso.

Sharon Timko is a public engagement specialist for the USDA Forest Service, and serves as the program manager for the National Collaboration Cadre.

Eli Typhina, Ph.D., currently works in the Forestry and Environmental Resources Department at North Carolina State University specializing in factors influencing human behavior within the natural world.

Sarah D. Upton is a Postdoctoral Fellow, Department of Communication, University of Texas at El Paso.

Gregg B. Walker is Professor, Department of Speech Communication, Oregon State University.

Kenneth C. C. Yang is Professor, Department of Communication, University of Texas at El Paso.

Index

public participation *(continued)* 141–42; civic technologies and, 9; critiques of traditional approaches, 4, 122, 124, 130, 177, 204–205, 303, 306; collaborative, collaboration movement in, 62–63, 71; constricted versus constructed potentiality, 228, 229, 230, 239, 242, 243; consummatory functions of, 154, 156; contested meanings-tyrannical or transformative, 203; critical for democracy, 203; Decide-Announce-Defend (DAD) approach, 5, 150, 151, 158, 162, 165; decision-space and, 56, 191–92; decision-support systems and, 28; definitions, 6–7, 19–20, 43, 63, 124, 125, 152–53, 175, 203, 206, 250, 277; deliberative inquiry and, 126; desire to be heard, 149; generative concepts and, 21, 32–34; ideal versus real, 2, 123, 177; infrastructure of, 5, 63–64, 195; interdisciplinary approaches to, 20–22; meaningful and consistent, 61–62; need for governance frameworks, 195; new forms of and innovations in, 5, 6, 8, 17, 19, 44, 195; normative, 203; paradox of (technical vs. lay public involvement), 49–50, 143, 176; policy dimensions, 63; power and, 176, 207; problems with generalization, 8–9, 17, 19, 29, 31, 194; public engagement and, 3, 6; public hearings—critique, 5, 10, 49, 150; radical potential of, 150, 152, 153, 165; role of communication, 22, 24, 25; scale of, 28; thick, thin, and conventional, 124, 127, 132, 135, 139, 143; traditional, 6, 44, 70, 152; trust and, 60–61; Women's

March and, 4. *See also* disruptive public participation, environmental public participation (EPP), indecorous voice, indecorum, social media, Web 2.0

public participation and space, 8; boundaries between private and public are elastic, 208; challenge of dispersed populations, 210–11; community engagement in informal, interstitial spaces, 220–21; communication geography, 205; formal and informal arenas, 204; importance of space in participatory design, 210; material and discursive dimensions, 204; science-society agoras, 208. *See also* public participation

public participation in local contexts, importance of local contexts, 7–8, 27, 30, 116; Boulder, Colorado, 141; Forest planning, 59; Indonesia, 221–22, 241; Merrimack River, 8, 29; New Zealand, 193; Southeast Asian community, 8, 22. *See also* public participation

public participation formats outside the United States, Africa-Environmental Impact Assessment (EIA), 48; China-Environmental Impact Assessment (EIA), 47–48; Denmark, 47; European Union, 49; Indonesia, 218–19; New Zealand, 182–87; Organization of American States, 49; United Kingdom, 47; United Nations-Aarhus Convention, 48–49; World Bank Institute, 47. *See also* public participation

public sphere, 3; contested, 207; Habermas ideal, 206–207. *See also* public participation